Intensive Biofloc Nursery System for Vannamei Shrimp

THE AUTHORS

S. Ezhilmathi has completed his M.F.Sc (Master of Fisheries Science) in Aquaculture from Tamil Nadu Dr.J. Jayalalithaa Fisheries University, Nagapattinam, Tamil Nadu. He won DSA – Gold medal for the best M.F.Sc thesis in Aquaculture, Dr. T.M. Samocha Gold medal for the best M.F.Sc thesis in Aquaculture, Best paper award in Agriculture scientific Tamil society (SCITSA), New Delhi, India and the TNJFU Appreciation Award for the Development of "NUTRIFLOC" Technology. He also participated in several national and international conferences. He authored four research articles, five semi-scientific articles and a book chapter on recent aquaculture techniques. He has also contributed as co-author in many training manuals prepared for the fish farmer's trainings organized by TNJFU.

Prof. S. Felix has 38 years of teaching and research experience in aquaculture. He is the Past President of World Aquaculture Society-Asian Pacific Chapter and he has also served as Director of the World Aquaculture Society(APC). Prof.S.Felix has served as Second Vice Chancellor of Tamil Nadu Dr. J. Jayalalithaa Fisheries University and also served as Dean of the Fisheries College in Tamilnadu, India.He has developed 40 Constituent units for the University including half a dozen new Fisheries Institutes to impart fisheries education and trainings. He developed and introduced the Raceway culture technology for the first time in India. Under his effective leadership the Aquatic Rainbow Technology Park and an Aquatic Rainbow Mall for ornamental fish culture were developed at Chennai, as first of its kind in the country. He has 10 patents published on various aquaculture technologies to his credit. He has published 15 Text Books, 50 Manuals and more than 150 Research papers and reports.. Introduced 30 new courses comprising UG, PG ,Ph.D and Paraprofessional programs. Implemented 30 R&D and Technology projects to develop a dozen aquaculture technologies for the benefit of farmers.

Intensive Biofloc Nursery System for Vannamei Shrimp

S. Ezhilmathi
S. Felix

2021
Scholars World
A Division of
Astral International Pvt. Ltd.
New Delhi – 110 002

Cataloging in Publication Data--DK

Courtesy: D.K. Agencies (P) Ltd. <docinfo@dkagencies.com>

Ezhilmathi, S., author.

Intensive biofloc nursery system for vannamei shrimp / S. Ezhilmathi, S. Felix.

pages cm

Includes bibliographical references.

ISBN 9789354616273

1. Shrimp culture--India--Tamil Nadu. 2. Whiteleg shrimp--India--Tamil Nadu. 3. Shrimps--Habitat--India--Tamil Nadu. I. Felix, S., author. II. Title.

LCC SH380.64.W45E94 2021 | DDC 639.58 23

Published by : **Scholars World**
A Division of
Astral International Pvt. Ltd.
– ISO 9001:2015 Certified Company –
4736/23, Ansari Road, Darya Ganj
New Delhi-110 002
Ph. 011-4354 9197, 2327 8134
E-mail: info@astralint.com
Website: www.astralint.com

Preface

Aquaculture is going to be an essential fish protein supplementing sector for the emerging global population. The aquaculture sector of India witnessed a boom after the introduction of White leg shrimps. Penaeus vannamei is the right choice for intensive and bio-secure closed grow-out culture. Due to limited natural resources, especially water and land, farmers shift to culture shrimps in intensive and super-intensive methods. High-density intensive nursery systems need high amounts of feed and other inputs to operate the systems. This will affect water quality due to the accumulation of organic wastes in the culture system, resulting in high production of ammonia in the system. Several nitrogen removal techniques have been developed to overcome the problems associated with ammonia accumulation in the aquaculture system. One such technique is Biofloc or Aerobic Microbial Floc(AMF) Technology. Biofloc technology can be defined as a technique based upon the development of heterotrophic bacteria within the culture system for zero or minimal water exchange. By culturing the P. vannamei with biofloc, protein utilization by the animal is nearly double and increases the digestive enzyme secretion and immune response of the cultured animals.

The book combined an outlining and updating of bioavailability assessment of crude protein and lipid in intensive biofloc nursery culture of P. vannamei. It provides the details on the effects of biofloc in feed utilization, floc characteristics, the bioavailability of nutrients, extracellular enzyme activity, nutrient absorption and growth performance of P. vannamei juveniles raised in indoor nursery raceway systems at high density in zero water exchange conditions. It is believed that this book will be useful to the students, teachers and researchers on the subject.

Authors

Contents

CHAPTER 1

Introduction

Aquaculture is the fastest growing sector, which provides an important and cheapest source of quality protein for human consumption. Global aquaculture production has grown dramatically in the last two decades and has become a significant sector of national development and poverty reduction plans in different regions of the world. The world production is about 167.2 million tons (FAO, 2016), and the demand for aquatic food keeps on increasing. The world per capita fish consumption increased from 9.9 kg in 1960 to 20 kg in 2014 (FAO, 2016). The aquaculture sector of India witnessed a boom with the introduction of White leg shrimps (*Litopenaeus vannamei;* currently known as *Penaeus vannamei*) in 2004. *P.vannamei* is a good choice for intensive and bio-secure closed grow-out culture. Indian seafood, especially frozen shrimps, has gained popularity in USA accounting for 30 % in terms of USD in 2017, European Union (18 %), and South East Asian countries (30%) over the last decade, it has supported the sector to grow at a faster pace due to rapid commercialization of *P.vannamei* shrimps. It is estimated that to supply aquatic food the production need to reach 196 million tons by 2025 (FAO, 2016). In view of the, limited natural resources, especially land and water, production of shrimp culture is being improved by adopting super intensive and intensive techniques. Super intensive system production of shrimp is gaining increased attention over worldwide to increase the aquaculture production by adopting nursery system between the hatchery and grow-out pond culture (Arnold et al., 2006; Correia et al., 2014) . High density intensive nursery systems need high amounts of feed and other inputs to operate the systems. This will affect water quality due to the high concentration of organic compound deterioration (Avnimelech, 2007). As fish biomass can assimilate 20–30% of feed and the remaining 70–80% of the feed will be accumulated in the water body (Avnimelech and Ritvo, 2003; Gross et al., 2000), resulting in high production of ammonia in the system. For overcoming ammonia accumulation in aquaculture system several nitrogen removal techniques have been developed. Such as Recirculating Aquaculture System, Periphyton based Aquaculture and Bio-floc technology or Aerobic Microbial Floc(AMF) Technology.

AMF in aquaculture can be defined as a technique of aquaculture system management, which is based upon the development of heterotrophic bacteria within the culture system for zero or minimal water exchange. AMF technique uses continuous aeration and mixing of organic matter in culture system. It will enhance the decomposition of organic matter under aerobic condition. The unused microbial protein will be used by the heterotrophic bacteria and converted into the microbial protein (Felix et al., 2015). The intensive nursery rearing of shrimp in limited or zero water exchange system helps to reduce the probability of crop loss and increase the production.

AMF has advantages like recycling of the nitrogen species within the culture system, reduction of nutrient input and also acts as an extra food source for the cultured animals. Protein utilization by the animal is nearly double and increases the digestive enzyme secretion and immune response of the cultured animals. By culturing the *P.vannamei* with AMF, growth and immune level of shrimp can be increased. Several studies performed on the use of biofloc as an *insitu* produced feed, and they indicate that biofloc can be taken up by aquaculture species and uptake depends on the species and feeding traits, animal size and floc density (Avnimelech, 2009).

AMF technology is applicable to both freshwater and seawater systems to control water quality and to produce as an additional *insitu* feed source. The potential feed gain on the application of biofloc technology is estimated about 10 –20 % (Schryver et al.,2008) with this, production costs will decline considerably since feed represents 40–50% of the total cost (Craig and Helfrich,2002).

Microbial flocs can be produced externally from shrimp tanks for treating a fish effluent. Microbial flocs produced in sequencing batch reactors could be dried and incorporated into a pelleted feed for shrimp. If this alternative feed proved to be successful, it could offer the shrimp industry to replace the marine animal meals and traditional plant meals.

The protein, lipid, carbohydrate and Vitamin C contained within the biofloc represent a considerable fraction of the nutritional requirements of several aquaculture species (Crab et al., 2010a; Tacon et al., 2002). The application of biofloc technology indirectly improved the feed conversion and protein retention, indicating that the consumption of biofloc helps in improving the growth of the animal.

The AMF based culture in shrimp aquaculture showed the increased growth rate due to the presence of minerals and extracellular particle in the floc. It also influences the digestive processes of the shrimp by increasing the exogenous and endogenous enzymes activity by altering the intestinal microflora balance. The biofloc also has a stimulating effect on the digestive enzyme activities of the shrimp.

As the nutritional properties of flocs and water quality dynamics are influenced by the type of carbon source used to produce the flocs. Different organic carbon sources stimulate specific bacteria, protozoa, and algae (Crab, 2010). However, further research focusing on the use of low cost non-conventional distillery industrial residues and agricultural wastes as a carbon source needs to be explored as carbon source in biofloc technology.

As extracellular enzymes, protein and lipid availability in the aerobic microbial floc enhance the growth rate and survival of aquatic animals in the culture systems. However, biofloc development and its relationship to bioavailability, nutrient absorption with the incorporation of biofloc as feed remain poorly understood.

The current study was therefore conducted to investigate the effects of *insitu* and *exsitu* biofloc feed utilization, bioavailability of nutrients, extracellular enzyme activity, nutrient absorption and growth performance of *P.vannamei* juveniles raised in indoor nursery raceway systems at high density in zero water exchange conditions.

Objectives

1. To assess the physical and chemical characteristics of Aerobic Microbial Floc produced in *insitu* and *exsitu* nursery system.
2. To trace extracellular and digestive enzymes activities in floc and shrimp in the culture system.
3. To study the zootechnical performance of shrimp based on the nutritional composition of AMF at the nursery phase.

CHAPTER – 2

Review of Literature

Fisheries and aquaculture remain important sources of food, nutrition, income and livelihoods for millions of people around the world. Aquatic fish production transits from capture fisheries to culture fisheries through increasing number of farmed species. A milestone was reached in 2014 when aquaculture sectors contributed to the supply of fish for the human consumption overtook that of wild caught fish for the first time (FAO, 2016). Aquaculture has been responsible for the impressive growth in the supply of fish for human consumption. Aquaculture as a food-producing sector is recognized for its ability to contribute to the global fish supply. Aquaculture production is projected to rise from 40 million tonnes by 2008 to 82 million tonnes in 2050 (FAO, 2016). The necessity to increase aquaculture production has been triggered by the increasing demand per capita in parallel to the increase of global population. However, the development of a sustainable aquaculture industry is particularly challenged by the limited availability of natural resources as well as the impact of the industry on the environment (Costa-Pierce et al., 2012; Verdegem and Fuchs, 2013). One possible way to increase the production and productivity is to intensify the system with more farm inputs restricted with limited discharge. Operating intensive culture systems with consistent production has been a major challenge for shrimp producers. Furthermore, outbreak of diseases in the early stage of the culture encouraged the farmers to adopt relatively small enclosed structures as nurseries for shrimp farming. Use of nursery facilities in areas affected by viral diseases can serve as a primary quarantine center, protecting the young shrimp from exposure for a longer period.

2.1 Nursery Rearing of shrimp

Nursery phase in shrimp farming have advantages over direct stocking, which increases control over stock inventories, water quality and feed management (Samocha and Benner, 2001). It provides larger and harder shrimp to be stocked and facilitates a higher number of crops per year as a shorter period is needed to reach a marketable size (Clifford,1985; Fast,1991). Use of nursery systems can also prevent the spread of diseases as post larvae can be kept under quarantine during the nursery. The advantages of two-stage production systems have been summarized by several authors (Fast,1991; Sandifer and Smith,1977; Samocha and Lawrence,1992; Felix.S,2006; Felix et al., 2015b). In general, two-stage shrimp production systems can improve PL inventory control and reduce losses to predation. Furthermore, since nursery tanks can be stocked at high densities, they can provide for more efficient utilization of the grow-out space with greater control over water quality and feed consumption. Furthermore, stocking these larger juveniles can also result in shorter production cycles and a higher number of crops per year (Clifford,1985; Brown and Briggs, 1991; Fast,1991, Peterson and Griffith,1999; Samocha *et al.*, 2001).

2.2 Shrimp nurseries in Raceways

Stunner et al. (1992) and Samocha et al. (1993) described the first nursery prototype system of six 40 m^3 raceways at the Texas Agri Life Research Mariculture laboratory. Cohen et al. (2005) reported good shrimp growth (1.1 g juveniles), high survival rate (>98%), low FCR (<1), and high yields (>4 kg/m^3), while using only low water exchange (1.1%/day). Keeping these shrimp in the nursery system for 50 days had no adverse effects on growth and survival in the grow-out phase at a density of 3,300 PL/m^3.

Samocha et al. (2004) provided a detailed description of the four 97.5 m^2 (147.6 m^3) greenhouse enclosed raceways that were used for nursery and grow-out of *P. vannamei* using low-salinity ground water (1.8–2.6 ppt). The Stocking and harvest data for post larvae of *Penaeus vannamei* reared in greenhouse enclosed raceways using low-salinity well water includes 19,200 PL/m^2 and 20,400 PL/m^2 with a 100 % survival on 35 day's nursery trial.

Intensive nursery rearing of shrimp in two 45-ton capacity greenhouse enclosed raceway system was studied by Chon et al. (2005) at a stocking density of 3300/m^2. Prangnell et al. (2016) reported the super intensive nursery culture of P. vannamei in 35 ton's capacity raceway with limited water exchange biofloc system at the stocking density of 457 PL/m^2 with 80% survival. Esparza-Leal et al. (2015) documented the nursery culture of Penaeus vannamei in RAS and biofloc system with different stocking density (1500,3000,6000 and 9000/m^3) and suggested to adopt optimum stocking density of 1500/m^3 to obtain higher growth rate and weight.

2.3 Aerobic microbial floc technology

The biofloc system was developed under the principle of waste water treatment, in which the microbes utilize the feces of the cultured organisms, transforming it into simple, complex organic products that can be used by other animal and return to the food cycle (Avnimelech and Kochba, 2009).

Biofloc are defined as aggregates (flocs) of algae, phytoplankton, free and attached bacteria, grazers of bacteria such as rotifers, nematodes, ciliates, flagellates protozoa and copepods, living and dead particulate organic matter such as faeces, uneaten feed, yeast, cyanobacteria, particle colloids, organic polymers, dead cells and inorganic particles (Burford *et al.,* 2003).

Aerobic microbial floc technology and biofloc technology falls in same line with continuous aeration and intensive stocking density. Under the continuous heavy aerated system, the activity of aerobic bacterial population could effectively utilize the nitrogen waste, and carbon in the system and converts them to microbial protein (Felix, 2012; Felix et al., 2015).

Biofloc Technology (BFT) is the technique of enhancing water quality through the addition of external carbon to the aquaculture system (Crab et al., 2012).

These microbial communities flocculate in the water column and feeds on organic substrates (high carbon, less nitrogen) and thus, forced to take up nitrogen along with the carbon from the water to produce the protein requirement for their cell growth and multiplication (Hargreaves, 2006). Simultaneously, unwanted components like ammonia will thus, be eliminated through the microbial uptake, convert them into microbial protein and ultimately into animal protein as fish harvest the biofloc (Ebeling et al., 2006; Schryver et al., 2008). For developing the flocs, Carbon: Nitrogen (C/N) ratio needs to be balanced in the system which can be done either by addition of extra carbon sources or by reducing feed protein inputs so that the carbon requirements for the heterotrophic bacterial biomass formation can be met (Avnimelech, 1999; Hargreaves, 2013).

By adding a carbon source to the culture medium in limited-discharge systems (i.e., by changing the C/N ratio), it is possible to obtain a significant enhancement of bacterial growth and of the fixation of toxic nitrogen metabolite species (Ebeling et al., 2006; Hari et al., 2006; Avnimelech, 2009; Crab, 2010). The bacteria transform toxic form of nitrogen (ammonia) to nitrate that is toxic only at high concentrations by the process called nitrification. The zero water exchange system thus tends to accumulate the nitrate in long run and hence usually the nitrate level in biofloc systems tends to increase. The waste nitrogen is repeatedly cycled between dissolved ammonia and solids of algae or bacteria (Hargreaves, 2013). The technology is efficient and environmental friendly as the nutrients are recycled and reused within the system itself (Emerenciano et al., 2013).

The advantages of the technology in aquaculture has been well documented which includes low feed and water input (economical), less risk of pathogen introduction and diseases, increased growth, survival and crop yield (Otoshi et al., 2009; Crab et al., 2009, Samocha et al., 2010; Krummenauer et al., 2011; Perez-Fuentes et al., 2016). It also lowered the feed conversion rate by utilizing the *insitu* natural feed and has small carbon foot prints, hence, reduced environmental impacts (Krummenauer et al., 2014). It is also robust, easy to operate and economically feasible (Crab et al., 2012). The zero water exchange system has advantages of maintaining temperature and heat fluctuations (Crab et al., 2009). It supports nitrogen removal even when organic matter load and biochemical oxygen demand of the system is high (Avnimelech, 2009).

2.4 Management of AMF

There are several factors that influence floc formation and floc structure. Ritvo et al. (2003) studied the effect of salinity and the pH on the colloidal properties of suspended particles in super intensive aquaculture systems. pH of 7.5 is found to be optimum for better flocculation in shrimp aquaculture (Jain and Srivastava, 2012). The mixing intensity imposed by a selected aeration device at a certain power input will determine the steady-state of the floc size. It is an equilibrium between the rate of aggregation and the rate of breakage in floc size distribution (Spicer and Pratsinis,1996; Chaignon et al., 2002). Dissolved oxygen also influences the biofloc formation. The organic carbon sources of choice to a large extent determine the composition of the floc produced (Mikkelsen et al.,1996; Hollender et al., 2002; Oehmen et al., 2004).

Lancelot and Billen. (1985) observed that when the C/N ratio of the organic matter is higher than 10, immobilization of inorganic nitrogen takes place. Hargreaves. (2006) suggested that if the organic matter is low in nitrogen content some of the N for the microbial growth must be obtained from the aquatic environment and become immobilized as microbial protein. Asaduzzaman et al. (2008) investigated the effect of C/N ratio in ponds with and without substrate addition for periphyton development on the production of *Macrobrachium rosenbergii*. Increased C/N ratio and substrate improve the heterotrophic bacterial abundance. A concentration of about 10 mg NH_4-N L^{-1} could almost completely be removed within 5 h after the addition of glucose at C/N ratio 10, and without the accumulation of nitrite and nitrate (Avnimelech,1999).

2.4.1 Distillery spent wash

It is one of the cheapest carbon source and unwanted waste residual liquid generated from alcohol production in distilleries. In India there are around 400 distilleries with total production capacity of about 3,800 million liters of alcohol (A.I.D.A, 2007). After alcohol distillation, large amount of a highly colored spent wash remain in the still to be disposed of, every one-liter alcohol produced from 10–15 liter of spent wash was generated (Chhaya and Kumar, 2014). This digested spent wash contains readily available carbon in spite of its total sugar content (Kumar et al 1998). Spent wash having a higher concentration of essential nutrients and minerals (Jain and Srivastava, 2012). Liu. W et al. (2017) reported that addition of DSW had a nitrogen dynamics process similar to that of the ASW substrate. Menaga et al. (2017) reported that microbial floc enriched DSW used as potential carbon without affecting the growth of shrimp.

2.5 *Insitu* biofloc culture

Avnimelech et al. (1986) stated that biofloc technology retains waste and converts into insitu food. Avnimelech, (1999) found 20% reduction in feed requirement, while culturing tilapia in pond, shrimp can utilize the biofloc and digest it as insitu food source which was available for 24 hours in the culture system. Hari et al. (2004) facilitated the development of heterotrophic bacteria and the related insitu protein synthesis by increasing the C/N ratio of the feed and further increasing the C/N ratio through carbohydrate addition to the ponds. The biofloc technology (BFT) is a sustainable technique used in aquaculture to maintain good insitu water quality through the development and control of dense heterotrophic microbial biofloc by adding carbohydrate to the water (Avnimelech et al., 1989; Avnimelech, 1999; Crab et al., 2007; Crab et al., 2009; Crab et al., 2010a).

Sinha et al. (2008) further reviewed the potential use of biofloc as an anti-inflammatory strategy in aquaculture and suggested insitu biofloc culture can be a novel strategy for disease management on a long term basis in contrast to conventional approaches such as antibiotic, probiotic and prebiotic applications. Hargreaves, (2013) reported shrimp indicates that insitu culture water contains growth enhancing factors such as microbial and animal proteins, that boost the growth of animal and floc were used as a supplemental food source for the grazers like shrimp or tilapia.

2.6 *Exsitu* biofloc culture

Kuhn et al. (2009) prepared exsitu biofloc feed by using sequencing batch reactors, these biofloc meal were used as ingredient for juvenile shrimp diet. On the other hand, biofloc meal was produced with activated sludge system by Neto et al. (2015), fed to shrimp at 30% replacement of fishmeal in diet. Hende et al, (2016) also reported the production of biofloc by using pikeperch culture water treated through a pilot reactor. He also included 80g/kg of the biofloc meal without deleterious effect on shrimp growth, survival and FCR. Bauer et al. (2012) produced a biofloc meal from an effluent of a super intensive shrimp farm. Izquierdo et al. (2006) have demonstrated that the nutritional contribution of the floc to shrimp in mesocosm culture reduces or eliminates the need for a dietary source of fish oil. Ju et al. (2008) collected biofloc from marine shrimp culture tanks as whole floc and with several extraction modes as a diet or diet component for shrimp. McIntosh et al. (2000) analyzed suspended flocs from commercial shrimp ponds in Belize managed as BFT systems. They found higher protein levels in suspended floc than in the feeds they were derived from the floc. Dried biofloc been proposed as an ingredient to replace fishmeal or soybean meal in aquafeed as biofloc meals are rich in nutrients while they are also free from deleterious levels of mycotoxins, anti-nutritional factors and other constituents that limit its use in aqua feeds (Emerenciano et al., 2013). Large-scale production of biofloc meal for use in aquaculture though would be hard at commercial level could result in environmental benefits to marine and coastal ecosystems, as the need for wild fish as an aquafeed ingredient will be reduced (Kuhn et al., 2010; Bauer et al., 2012). Large scale production of biofloc meal for use in aquaculture at commercial scale result in the environmental benefits to marine and coastal ecosystems, as the need for wild fish as an ingredient will be reduced (Kuhn et al., 2010; Bauer et al., 2012).

2.7 AMF as a feed for different aquaculture species

Under the concept of using biofloc as an insitu produced feed, several studies were performed and they indicate that biofloc can be taken up by aquaculture species depending on the species and feeding traits, fish size, floc size and floc density (Avnimelech, 2007; Crab et al., 2009; Crab et al., 2010b; Crab et al., 2010a). In this work Macrobrachium rosenbergii, Litopenaeus vannamei and tilapia (Oreochromis niloticus x Oreochromis aureus) were all positively identified to take up biofloc and profit from this additional protein source. This makes biofloc technology applicable to both freshwater and seawater systems to control water quality and as a surplus through insitu feed production. The potential gain on feed is estimated to be in the order of 10-20% (Schryver et al., 2008). With this, production costs will decline considerably since food represents 40-50% of the total production costs (Craig and Helfrich, 2002).

Neto et al. (2015) fed the shrimp with 30% replacement in the feed but the microbial floc was produced from clear sea water gave better growth in shrimp culture. Khun et al. (2009) succeeded in replacing soybean meal and fish meal in shrimp feed with biofloc produced with tilapia farm effluents. Khun et al. (2010) reported as the 10 to 30 % biofloc inclusion in the juvenile diet of shrimp showed a better growth and survival.

In an experimental trial, Bauer et al. (2012) substituted the fishmeal with soybean concentrate and biofloc meal and did not found a negative effect on the productive parameters of the Pacific white shrimp. Nunes et al. (2010) used an experimental feed (less than 250 g kg-1 crude protein; mix of a commercial diet and commercial poultry feed) in a biofloc technology (BFT), demonstrating that the Pacific white shrimp (P. vannamei) had a better performance in the BFT system compared with a traditional diet (370 g kg-1 crude protein). Valle et al. (2015) evaluated the effect of the gradual replacement (100, 200, 300 and 400 g kg-1) of fishmeal by fish protein hydrolysate and biofloc flour in P. vannamei post larvae and suggested to replace fishmeal by either of these two ingredients.

2.8 Nutritional profile of AMF

2.8.1 Proximate composition

Bureauet al., (2004) noted the main factors affecting nitrogen retention by animal was amino acid composition (the concentration and profile), and the balance between digestible protein and digestible energy. Crab et al. (2010a) reported that biofloc contain adequate protein, lipid, carbohydrate and ash required for an aquafeed. Emerenciano et al. (2013), reported that the dry-weight protein content of biofloc ranges from 25 to 50%, fat content ranges from 0.5 to 15% and ash content ranges from 13-46%. A narrative compilation of proximate composition of biofloc is given in the table 2.1.

2.8.2 Fatty acid and amino acid profiling of AMF

Biofloc has a good fatty acid profile to satisfy the shrimp nutritional requirements (Morenoarias et al., 2018). Fatty acid profile culture of shrimp was not affected by the different fishmeal content in the feed (Lim et al., 1997). According to Ekasari et al. (2010) biofloc were rich in the 16:0, 16:1 n-7, and 18:1 n-7 fatty acid. The total n-3 PUFAs ranged from the 0.6–0.8 mg/g of dry weight, which are higher than the bacterial based microbial community (Izquierdo et al., 2006). Emerenciano et al. (2013) suggested that the fatty acid profiles may vary based on the carbon sources employed, the type of water used (freshwater, seawater) and the methods of production. The proximate composition of biofloc derived from different carbon source by several authors are given in table 2.1.

Table 2.1 Proximate analysis of biofloc reported by various authors.

Sl.No.	References	Protein (%)	Lipid (%)	Ash (%)
1	McIntosh et al. 2000	43	12.5	26.5
2	Cuzon et al. 2004	27.3%–31.6%	3.7-4.2	-
3	Soares et al. 2004	12-42	2-8	22-46
4	Wasielesky et al. 2006	31.1	0.5	44.8
5	Ju et al. 2008a	26-41.9	1.2-2.3	18.3-40.7
6	Ju et al. 2008b	30.4	1.9	38.9
7	Azim and Little, 2008	38	3	12
8	Kuhn et al. 2009	49	1-1.3	13.4

Sl.No.	References	Protein (%)	Lipid (%)	Ash (%)
9	Kuhn et al. 2010	38.8	<0.1	24.7
10	Maica et al. 2012	28.8-43.1	2.1-3.6	22.1-42.9
11	Emerenciano et al. 2012	30.4	0.5	39.2
12	Emerenciano et al. 2013	18.2-29.3	0.4-0.7	43.7-51.8
13	Anand et al. 2013	24.4	3.53	31.98
14	Ekasari et al. 2014	17.2-27.8	6.0-7.5	-
15	Hende et al. 2014	15.8 – 27.7	2.64 – 4.40	50.3-70.2

The presence of essential fatty acids such as Linoleic acid (C18:2 n-6), Linolenic acid (C18:3 n-3), Arachidonic acid (C20:4 n-6), Eicosapentanoic acid (C20:5 n-3) and Docosahexaenoic acid (C22:6 n-3) in biofloc.

Ju et al. (2008a) demonstrated that the concentration of free amino acids such as alanine, glutamate, arginine and glycine, which are known attractants in shrimp diet (Nunes et al., 2006), are present in biofloc. Levels in biofloc were found to be comparable to that of the shrimp commercial diet suggesting that biofloc are likely to be recognized as food particles by some aquaculture organisms.

Biofloc also contain various bioactive compounds including essential fatty acids, carotenoids, free amino acids and chlorophylls (Ju et al., 2008b), trace minerals (Tacon et al., 2002) and vitamin C (Crab et al., 2012) which are known to have positive effects on aquaculture animals including the enhancement of antioxidant status, growth, reproduction and immune response.

2.8.3 Bioavailability of AMF

Silva et al. (2013) reported that the application of biofloc technology on Pacific white shrimp super intensive culture considerably enhanced N and P utilization efficiency up to 70% and 66%, respectively, relative to conventional intensive culture systems with regular water exchange. The apparent digestibility coefficient and bioavailability of nutrient were directly proportional and decrease in ADC will lead to decrease in the bioavailability of nutrients. In general, the ADC of crude protein in the feedstuff used in shrimp dies very between 58–97% (NRC, 2011), these includes anchovy meal (ADC 83–89%), soybean meal (ADC 89–97%) etc., The bioavailability of biofloc meal was first reported by Neto et al. (2015) and found that biofloc meal incorporated feed has the ADC of 78–85%.

Protein quality of a diet for aquaculture species is determined by the amino acid composition of the protein and the bioavailability of the amino acids present. Biofloc may contribute to the supply of essential nutrients and digestive enzymes either through the stimulation of endogenous production or microbial secretion (Xu and Pan, 2012; Anand et al., 2014), and the enhancement of nutrient bioavailability that facilitates higher nutrient assimilation. Most of the studies applying biofloc technology confirmed that the N and P waste in this system could be reduced, corroborating the role of this system on the improvement of aquaculture productivity and the reduction of environmental impact from aquaculture unit among others, (Perez-Fuentes et al., 2016; Luo et al., 2014).The consumption of the microbial biomass in the biofloc might therefore facilitate P assimilation, in particular the indigestible one, from the feed to the cultivated organisms thus reducing the nutrient waste (Luo et al., 2014; Silva et al., 2013).

2.9 Extracellular enzyme activity of AMF

Moss et al. (2000) reported microbial floc and attached organisms could exert a positive effect on the digestive enzyme activity of shrimp. Xu et al. (2012) proved as microbial enzymes could help in the breakdown of the protein, carbohydrate, and other nutritional ingredients of the feed into smaller units, promote microbial floc could presumably facilitate feed digestibility and absorption. Furthermore, different bacteria can produce the different enzymes in the biofloc (Yu et al., 2008). Biofloc not only provide the insitu feed but also produce extracellular enzymes which can facilitate feed digestion and utilization (Xu et al., 2012; Xu and Pan, 2013). Wang et al., (2015) found that based on the carbon source, the extracellular enzyme activity also differs and recorded highest extracellular enzyme activity in 60% molasses, 20% wheat flour and 20% corn flour as a carbon source.

Many enzymes such as aspartate aminotransferase (AST) and alanine aminotransferase (ALT) take part in amino acid metabolism within hepatopancreas (Pan et al., 2003) is found higher in biofloc supplemented feed. Enzymes involved in energy production cycle like malate dehydrogenase (MDH) and lactate dehydrogenase (LDH) also found to be higher in biofloc based culture system (Anand et., 2014). The enzyme glucose-6-phosphate dehydrogenase (G6PDH) is involved in synthesis of fatty acids, cholesterol and sphingolipids (McWhinnie and Kirchenberg, 1962). This enzyme activity increased with phagocytosis indicating better oxidative status of the animal in biofloc based culture (Anand et la., 2017)

2.10 Digestive enzyme activity of AMF

Xu and Pan, (2012) reported the enhancement of protease and amylase activities of the shrimp grown in the biofloc based system and suggested that appropriate extracellular enzyme activities are present in biofloc. Few beneficial bacteria such as Bacillus sp. in the ingested bio floc could facilitate the modification of physiological and immunological status of the host, through the colonization in the gastrointestinal tract (Zhao et al., 2012; Xu and Pan, 2013). Xu et al. (2012) suggested that the microbial flocs could exert a positive effect on the digestive enzyme activity of shrimp which can facilitate feed digestion and utilization. In biofloc system different microorganisms can produce different kind of enzyme to breakdown the protein, lipid and other partials, when these particles ingested in shrimp it will also improve the digestion of the animal. Wang et al. (2015) reported protease enzyme activity was higher in the biofloc cultured shrimp compared to the clear water system.

2.11 Physical and chemical characteristic of AMF

AMF, also known as biological or bacteria aggregated with different group of microorganisms was constructed by bioflocculation (Keiding and Nielsen, 1997). It is accomplished by chemical induction, interaction and effect of polyelectrolytes. Even small changes in the ionic strength and ion composition of the water can have substantial influence on the structural properties of flocs. Particularly Ca^{2+} has been shown to be a significant factor in floc formation (Keiding and Nielsen, 1997). In addition, the presence of calcium ions also seems beneficial in the protection of fish species against heavy metal toxicity (Abdel-Tawwab et al., 2007). Bioflocculation may also be steered by the choice of organic compound used as food for the biofloc.

The surface of a bacterium surrounded by polymeric compounds is in general negatively charged (Zita and Hermansson, 1994). The negative charge of flocs lies within the range of -0.2 to - 0.6 meq/g with a zeta-potential of -20 to -30 mV was reported by Liu and Fang, (2003).

The increased occurrence of granule filaments of floc was observed by Schryver and Verstraete, (2009) during shifts in the C/N ratio. Similar differences in the morphological structure of the microbial biomass types were observed by using different as carbon source Tay et al. (2001). McSwain et al. (2005) summarized that in addition to carbon type, the DO concentrations, shear force, settling time and even feast/famine regimes influence granule formation and morphology. McSwain et al. (2005) stated that floc-forming bacteria had a selective advantage (regarding to substrate-uptake kinetics and starvation sensitivity) over filamentous organisms at short feast relative to long famine periods (1:4).

Martins et al. (2003) reported increased amounts of filamentous bacteria in flocs compared to floc formers at DO levels less than or equal to 1.1 mg O_2 L^{-1}. Increased aeration would not only increase DO levels but also the shear force applied to the flocs (Tay et al., 2001). Moreover, the quality of the bio-flocs in terms of protein content was similar to fish feed while the accumulation of PHB can represent a major added value. According to Crab et al. (2010), FVI should be higher than 200 mL/g to avoid the flocs from settling too fast in regions of lower turbulence.

Since changes in floc density are usually very small, a concept of effective density, i.e., the buoyant density, defined by Magara et al. (2016) and Tambo and Watanabe, (1979) as has been used in most floc density studies. The results from almost all the studies of the size-density relationship for inorganic or biological flocs showed that the floc effective density would decrease as the floc size increased (Magara et al., 1976, Mitani et al., 1983, Li et al., 1986, Boadway 1978).

Filamentous bacteria have an advantage over non-filamentous bacteria at low substrate levels due to their higher surface-to-volume ratio. Moreover, the filaments can penetrate outside the flocs and thus are exposed to higher substrate concentrations than the non-filamentous bacteria that mainly grow within the flocs (Martins et al., 2003).

The nutritional composition of floc is determined by the nitrogen recovery, biofloc of 48 μm and 100 μm particle sizes could meet the requirement of the shrimp (Ekasari et al., 2014). Furthermore, in an aquaculture system with biofloc technology application, floc size may also relate to the uptake potential of the biofloc by the cultured organism, the digestibility of the flocs as well as the nutritional value of the flocs (Schryver et al., 2008). The measurement of individual floc-settling velocity by photographic methods have been applied by several investigators (Magara et al, 1976; Tambo and Watanabe, 1979; Hermanowicz and Ganczarczyk, 1983).

Biggs and Lant, (2000) showed in case of activated sludge that for an average velocity gradient (or G-value) of 19.4 s^{-1}, the stable floc size was 130 μm whereas this was decreased in biofloc to 20 μm for a velocity gradient of 346 s^{-1}. For BFT, the steady-state floc size is an important feature as it has already been shown that the quality of food for different aquaculture species is also dependent on the food size (Garatun-Tjeldsto et al., 2006; Knights, 1983). In order to represent a nutrition source, the food particle size in case of e.g. cod larvae and Macrobrachium rosenbergii larvae required floc size in 250–1200 μm (Barros and Valenti, 2003).

Mean floc pore radius and tortuosity are the characteristics describing the fine floc physical structure. Although the porosity of flocs is clearly non-homogeneous, most models for floc and biofilms assume that the porosity is constant (Scuras et al., 1998). The porosity variations are especially important in relation to mass transfer, since any porosity gradient will result in the mass (substrate or oxygen) concentration gradients inside a floc. Several factors, besides structure, affect the size of flocs. Flocculi form during a fast mix, when the shear rate, defined by liquid velocity gradient, is the highest, whereas floc aggregates are present when the shear rate is low. The three different levels of aggregation - flocculi, microflocs and floc aggregate have different resistance to shear or strength (Li and Ganczarczyk, 1986). The strength of flocs can be determined from the relationship between the floc size and the mixing energy input in the flocculation experiment.

The porosity reported for activated sludge flocs increased at two different rates in a range from 43 to 99.6%. The porosity-increasing rate of smaller flocs (less than about 200 μm) was much higher than that of the larger flocs.

The dehydrogenase activity is based on the principle that dehydrogenase enzymes are produced by all living cells and the extent to which this enzyme group oxidizes organic matter can be related to the number of live cells present (Mahmoud and Ghaly, 2004) This group of enzymes transports electrons and hydrogen through a chain of intermediate electron carriers to a final electron acceptor (oxygen), resulting in the formation of water (Ghaly et al., 1989). The increase in dehydrogenase activity found to be increase in oxidation of organic nutrients in the biofloc systems.

CHAPTER – 3

Materials and Methods

3.1 Site of the experiment

The experiment was conducted over a period of 30 days in indoor nursery raceways in Brackish Water Research Farm Facility, TNJFU- OMR Campus, off campus facility of Dr. MGR. Fisheries College and Research Institute, Ponneri, TNJFU.

3.2 Experimental design

A completely randomized design (CRD) was followed for the experiments. The experiment was designed with one control and two treatments. They are:(1) culture in clear water with commercial feed supply (control); (2) culture in AMF system plus a commercial feed supply (*Insitu* or T1); and (3) culture in AMF system with floc based feed supply (*Exsitu* or T2). The commercial and AMF incorporated shrimp feed contain similar level of crude protein (35%). The experiment was carried out in larger commercial culture system.

3.3 Statistical analysis

All Statistical analyses were performed using SPSS 24.0 for Windows. One-way analysis of variance(ANOVA) was performed to examine the difference in growth parameters, water quality parameters, proximate composition and nutritional quality of Aerobic Microbial Floc, among the treatments. All mean separations were carried out by Duncan's Multiple Range Test at the significance level of $P < 0.05$. Two-way ANOVA was carried out for digestive enzyme assay at the significant of level of $P<0.05$ for separation of mean.

3.4 Experimental setup

3.4.1 Configuration of nursery raceways

The raceways (30-ton capacity) were operated with high-pressure rapid sand filters (20,000 LPH). Each raceway had a central longitudinal partition placed over a 5.1 cm (2 in.) schedule 40 mm dia PVC pipe designed to provide oxygen-rich water across the bottom of the raceway. Water oxygenation in each raceway was generated by two 5 hp twin-lobe air blower. This setup increased dissolved oxygen levels of the water going back into the raceway by mixing it with air or pure oxygen. In addition, each raceway was provided with eighteen 5.1 cm airlift pumps arranged in three equidistance banks on each side of the partition.

3.4.2 Water source and treatment

Borewell water of 23 ppt salinity was stored in 2000 L reservoir sump. The borewater was allowed to settle down for a week. Meanwhile, the water was allowed to be aged at least 78 hours within the reservoir ponds. This process is to ensure that the screening, chemical treatment and aging process are efficiently used before stocking in raceways. The water was chlorinated with a commercial grade calcium hypochlorite (Bleaching powder 33% of active chlorine) to achieve an initial chlorine concentration of 20 ppm with targeted residual chlorine of 1 ppm after 38 hrs. After dechlorination, with sufficient aeration the salinity was adjusted for 20 ppt with freshwater. The diluted water is directed to the sand filter for further refinement and it was filled in the raceways.

3.5 Aerobic microbial floc (AMF) production

3.5.1 *Insitu* treatment

The treated water was added with the following fertilizers for the development of AMF by following Taw N et al. (2006) protocol with slight modifications to promote heterotrophic condition for the culture. The quantity of fertilizers added for AMF development in indoor raceways were presented in table 3.1.

Table 3.1 Quantity of Fertilizers added for AMF development.

Day	Fertilizers	Quantity (g/ton)
1	Urea	1.1
	Triple Super Phosphate (TSP)	0.14
	Grain pellet	4
	Dolomite	7
2	Grain pellet	4
	Dolomite	7
3	Grain pellet	4
	Dolomite	7
4	Grain pellet	4
	DSW	10.1
5	Grain pellet	4
6	DSW	7

3.5.2 Distillery spent wash as novel carbon source

Distillery spent wash was collected from M/s. Rajshree Biosolution Pvt. Ltd situated at Coimbatore district, Tamilnadu, India. The physico-chemical characteristics such as pH, TDS, TSS and BOD were analyzed by Standard methods (APHA, 2005) in lab (Table.3.2). Temperature and pH of the spent wash were measured and kept in a controlled environment. Spent wash was used as the carbon source by adding it directly in the culture system based on the carbon-nitrogen ratio. In both AMF treatments 10:1 C/N ratio was maintained throughout culture period to regulate the heterotrophic bacterial growth (Avnimelech,1999).

Table 3.2 Physico-Chemical properties of Distillery Spent Wash.

Parameters	Results (mg/l)
Color	Dark Brown Colored Liquid
pH @ 25ºC	6.25
Total Chlorides	7729
Total Suspended Solids	42730
Total Phosphorus	66.6
Total Solids	326980
Calcium	6072.1
Magnesium	4421.4
Potassium	23055
Carbonates	Nil
Electrical Conductivity Micro ohms/cm @ 25ºC	112652
Total Organic Carbon	26%
Total Nitrogen	834
Organic Solids	252340
Inorganic Solids	74640

3.5.3 Calculation of carbohydrate requirement

The amount of carbohydrate addition required to reduce the ammonia was calculated according to Avnimelech, (1999).

$$\Delta C_{mic} = \Delta CH \text{ X } \%C \text{ X } E \quad (1)$$

Where ΔC_{mic} is the amount of carbon assimilated by microorganisms, %C is the carbon content of the added carbon source. The amount of nitrogen need for the production of new cells material (ΔN) depends on the C/N ratio on the microbial biomass which is about

$$\Delta N = \Delta C_{mic} / (C/N)_{mic} = \Delta \text{ CH X } \%C \text{ E} / (C/N) \quad (2)$$

Equation of (2) can be modified as

$$\Delta CH = \Delta N \text{ X } (C/N)_{mic} / (\%C \text{ X } E) \quad (3)$$

Carbon content in Spent wash (%C) 26%, microbial convention coefficient (E) approximately 0.4, $(C/N)_{mic}$ will be maintained at 10. Equation (3) converted to

$$\Delta CH = \Delta N \text{ X } 10 / (0.26 \text{ X } 0.4)$$

Based on the above equation addition of carbon source to AMF treatments was carried out.

3.5.4 *Exsitu* feed formulation

AMF meal was produced in the process of treating effluent from the shrimp culture system by using spent wash as a carbon source, this effluent was pumped to an aerated well mixed raceway (30 ton). Atmospheric air and water was used to aerate and agitate the water to make the AMF in suspension state. Carbon was added every-day at C/N ratio of 15:1, to help the microbes to reduce the organic matter and assimilate the nitrogen waste into microbial protein. In the beginning, 10 liter of AMF concentrate was added as an inoculum in raceways. When the floc volume reaches 30ml/l, once in two days AMF was harvested from the raceway with the help of 20μ mesh size hand net and dried under shadow for 12 hours. Ten kg of dried biofloc meal produced, was powered into fine particles and stored in air tight containers. The major shrimp feed ingredients used for *exsitu* feed preparation include: dried biofloc meal, squid meal (Blue lines Pvt Ltd), soy lecithin (NRC Pvt Ltd, Madurai), wheat flour, soybean meal, vitamins, minerals and fish oil.

All the ingredients were mixed with water to make a dough. The dough was steam cooked under pressure cooker for 15 minutes at 15 psi. After cooking, dough was pressed through a pelletizer with 1 mm die and dried in sunlight till the preferred moisture level (8–10%) was reduced. The prepared feed was stored in an air tight container. The ingredient composition of the experimental feed was shown in the Table 3.3. The fish meal replaced shrimp diet developed by incorporating AMF meal at 25% inclusion level was named under the brand name of 'NUTRIFLOC' by Fisheries College and Research Institute, Ponneri.

Table 3.3 Ingredient composition of formulated biofloc feed.

S. No	Ingredients	Inclusion level (%)	Inclusion level (kg)
1.	Biofloc mael	25	10
2.	Squid meal	22.5	9
3.	Soybean meal	22.5	9
4.	Yeast	0.625	0.25
5.	Wheat	22.5	9
6.	Fish oil	1.875	0.75
7.	Soy lechitin	3.75	1.5
8.	Trace minerals premix	0.625	0.25
9.	Vitamin premix	0.625	0.25
10	Total	100	40

3.5.5 Feed and feeding management

Avanti manufactured crumble feeds, (0.5-1mm) containing 35 % crude protein were used throughout the experiment (Correia et al., 2014) for *insitu* and control treatments. Biofloc based feed formulated with the complete replacement of fish meal holding 35% crude protein was used for the *exsitu* treatment. Feeding rates were based on observation of feeding behavior of the shrimps during the first few days and fixed at 20 % of the total stocked biomass, then adjusted weekly after every sampling. Daily feed rations were split into four equal amounts given at 6:00,10:00,18:00, and 20:00 hours to all the experimental groups.

3.6 Stocking of shrimp seed

PL 15 shrimp *(Penaeus vannamei),* were used for the experiments. Shrimp PL procured from Aqua Nova shrimp hatchery Kanathur, OMR Road, Chennai. The seed samples were tested for WSSV, AHPND and EHP infection three day's prior to stocking following two-steps PCR (3.4). The healthy shrimp were stocked in the raceway tanks with proper acclimatization, the at the rate of 1500 PL/m^3 raceways. The trail was conducted for a period of 30 days in raceways.

Table 3.4 Screening results of *Penaeus vannamei* PL for viral diseases.

S.No		Test	Raceways
1	Microscopic	Stress test survival in Formalin	80%
2		Salinity test	80%
3		Size variation	10%
4		Necrosis	10%
5		Muscle gut ratio	3:01
1	PCR	MBV	Negative
2		BP	Negative
3		HPV	Negative
4		IHNNV	Negative
5		TSV	Negative
6		HPV	Negative
7		WSSV	Negative
8		EMS /AHPND	Negative
9		EHP	Negative

MBV-Monodon Baculo Virus: **BP**-*Baculovirus penaei*: **HPV**-Hepatopancreatic parvovirus: **IHNNV**-Infectious Hypodermal and Haematopoietic Necrosis Virus: **TSV**-Taura Syndrome Virus: **WSSV**-White Spot Syndrome Virus: **AHPND**-Acute Hepatopancreatic Necrosis Disease: **EHP**- *Enterocytozoon hepatopenaei*

Plate – 1 AMF meal production in raceway

1A - Inoculum

1B – Spent wash

1 C – Harvesting AMF from raceway

1D – Harvested wet AMF

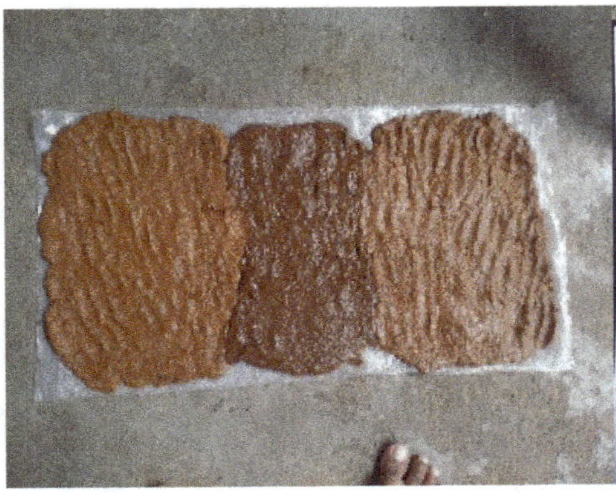

1 E – Dehydrating of AMF under sunlight

1 F – Dried AMF meal

Plate – 2 – Exsitu feed production

2A - AMF Feed ingredients

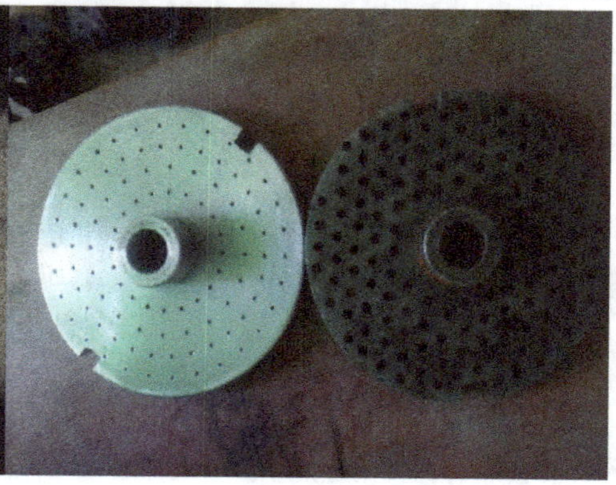

2B – 1mm die used for feed preparation

2 C – Assembly of feed pelletizer

2D – 1mm size AMF feed

2 E – Dehydrating of AMF feed under sunlight

2F – Commercial feed and AMF feed

3.7 Media, chemicals and Glassware's

Laboratory grade media, chemicals and glassware's were used. The glassware used in molecular and microbiological work, were sterilized in hot air oven (DTC-96, Expo Everflow, Chennai) at 180°C for two hours. The media, pipette tips *etc* were sterilized by autoclaving at 121°C for 15 minutes

3.8 Water quality

Water samples were collected from the experimental raceways during morning hours between 8.00 and 10.00 am for the whole experimental period. Visual analysis of water quality such as color and flocculation were observed daily. Temperature, salinity, floc volume and pH were measured in the experimental unit itself. Other parameters such as Dissolved oxygen (DO), Total alkalinity, Total hardness, Nitrate-Nitrogen, Nitrite-Nitrogen, Ammonia-Nitrogen, BOD, COD, Calcium and Magnesium were carried out in the lab following standard protocols (APHA, 2005).

Water pH was measured using portable pH meter (Deluxe pH meter 101) and salinity was measured using a hand-held refractometer (RHS 10 ATC). The dissolved oxygen content of the water was estimated using Winkler's titrimetric method (APHA, 2005).

Total Alkalinity and Total hardness of the water were estimated titrimetrically following the standard methods (APHA, 2005) using phenolphthalein, methyl orange and Eriochrome black-T as indicators respectively. Total Ammonia-N, Nitrite-Nitrogen Nitrate-Nitrogen, and Orthophosphate of the water samples were estimated using standard protocols (APHA, 2005).

Calcium content of the water was estimated titrimetrically following the standard methods (APHA, 2005) using Murexenic acid as indicators. Magnesium content of the water was estimated titrimetrically by following the standard methods (APHA, 2005) using Eriochrome black T as indicators.

Five days BOD was measured throughout the culture, initial and final DO was determined by using Winkler's titrimetric method (APHA, 2005). COD was estimated by permanganate method (APHA, 2005)

3.8.1 Microbiological parameters

For the microbiological study, samples were collected once in 6 days from culture water, 10 ml of water samples were collected from each tanks to enumerate the total plate count and Total vibrio count were estimated following standard protocol (APHA 2005).

CFU= average count / (volume of the sample X dilution factor)

3.9 Growth analysis

3.9.1 Percentage weight gain

Percentage weight gain was calculated using the following formula:

$$\text{Percentage weight gain} = \frac{\text{Final weight (g)} - \text{Initial weight (g)}}{\text{Initial weight (g)}} \times 100$$

3.9.2 Specific growth rate

$$\text{Specific growth rate, SGR (\%)} = \frac{(\ln \text{final weight} - \ln \text{initial weight})}{\text{Rearing period (day)}} \times 100$$

3.9.3 Feed convention ratio

Feed conversion ratio was calculated by the following formula-

$$\text{Feed Conversion Rate} = \frac{\text{Feed given (dry weight)}}{\text{Body weight gain (wet weight)}} \times 100$$

3.9.4 Feed efficiency ratio

The feed efficiency ratio was calculated by the following formula

$$\text{Feed Efficiency Ratio} = \frac{\text{Body weight gain (wet weight)}}{\text{Feed given (dry weight)}} \times 100$$

3.9.5 Protein efficiency ratio

The protein efficiency ratio was calculated by the following formula

Protein Efficiency Ratio = Body weight gain (wet weight) / Crude Protein fed

3.9.6 Survival rate

Survival was expressed as the percentage of live shrimps harvested from the final day of the expriment relatively to the total number initially stocked shrimp.

$$\text{Survival (\%)} = \frac{\text{Total number of animal harvested}}{\text{Total number stocked}} \times 100$$

3.9.7 Quantification of DNA and RNA ratio in muscle

Quantitative determination of nucleic acids in tissue was done by Pentose analysis following the method of Schneider, (1957). A 20% of homogenate was prepared with sucrose (0.25M) solution. Muscle of the shrimps were removed aseptically and 1 g of the muscle sample was taken and homogenized with chilled sucrose solution in a glass tube using a tissue homogenizer. For the removal of acid soluble compounds, 1ml of 20% muscle homogenate was mixed with 2.5 ml cold 10% TCA and centrifuged at 5000 rpm for 10 min. The supernatant was discarded and repeated the same step twice. Final sediment was extracted twice with 95% ethanol at 5000 rpm for 10 min. The lipid free sediment was then suspended in 1.3 ml of distilled water and 1.3 ml of 10% TCA. The mixture was then heated for 15 minutes at 90°C. It was centrifuged at 5000 rpm for 10 minutes and the supernatant was taken for further analysis.

Estimation of DNA was done by Diphenylamine method and RNA by Orcinol reagent method. One ml supernatant and 2 ml Diphenylamine reagent were mixed and heated for 10 minutes in boiling water. The optical density of the solution was taken at 600 nm.

For quantification of RNA in nucleic acid extract, 0.2 ml solution was taken and diluted with1.5 ml of distilled water. The mixture was heated with 1.5 ml of Orcinol reagent for 20 min in boiling water. The OD reading was recorded at 660 nm. The DNA content of the nucleic acid extract was calculated by the following equation:

$$\mu g \text{ DNA/ml} = \text{OD at 600 nm}/0.019$$

The RNA content of the nucleic acid extract was calculated by the following equation:

$$\mu g \text{ RNA/ml} = [(\text{OD at 660 nm} + 0.008) - (\mu g \text{ DNA/ml} \times 0.013)] / 0.116$$

3.9.8 Proximate analysis

Proximate analysis of the culture animal was done following standard methods (AOAC, 1995).

3.9.9 Moisture

The moisture content of the shrimp whole body, feed and floc were determined by taking a known weight of the sample in the petridish and drying it in hot air oven at 100–105°C till a constant weight was achieved. The difference in weight of the sample gave the moisture content, which was calculated using the following formula

$$\text{Moisture (\%)} = \frac{(\text{Wet weight of sample} - \text{Dry weight of sample})}{\text{Wet weight of sample}} \times 100$$

3.9.10 Crude protein

The nitrogen content of the tissue samples, feed and floc were estimated quantitatively by Kjeltec semi-automated method (2200 Kjeltec Auto Distillation, Toss Teccator, Sweden). The crude protein percentage was obtained by multiplying the nitrogen percentage by a factor of 6.25.

$$\text{Crude protein (\%)} = N_2 \text{ (\%)} \times 6.25$$

3.9.11 Ether extract

Ether extract of dried whole body samples, feed and floc were estimated by Soxhlet apparatus using petroleum ether (Boiling point 55 ± 5°C) as the solvent. The calculation was made as follows.

$$\text{Ether Extract (\%)} = \frac{\text{Weight of ether extract}}{\text{Weight of the sample}} \times 100$$

3.9.12 Ash

Ash content of the shrimp whole body, feed and floc were estimated by taking a known weight of dried samples in a silica crucible and placing it in a muffle furnace at 600°C for 6 hrs. The calculation was done as follows:

$$\text{Ash \%} = \frac{\text{Weight of ash}}{\text{weight of the sample}} \times 100$$

3.9.13 Organic matter

Organic matter of the shrimp whole body, feed and floc were calculated by the formula

$$OM\ (\%) = 100 - \%\ ash$$

3.9.14 Gross energy

Gross energy of AMF and shrimp samples, feed and floc were determined by using bomb calorimeter (Pan calorimetric Thermometer) energy of the dried material was determined by placing a weighted sample in the ignition cup of the bomb. Oxygen was filled at 25 psi and the bomb immersed in 2 liters of water. The samples were ignited and temperature increase of the water was recorded. Gross energy was calculated and expressed at Kcal g^{-1} of dry matter.

3.9.15 Total carbohydrate

The total carbohydrate (TC) of the shrimp whole body, feed and floc were calculated by subtracting the percentage of other nutrients from 100 (Hasting, 1969).

$$TC\ (\%) = 100 - \text{Crude protein}\ (\%) - \text{Ether extract}\ (\%) - \text{Ash}\ (\%)$$

3.9.16 Digestibility Energy

Digestible energy of the shrimp whole body, feed and floc were calculated as per standard physiological values (Halver, 1976) according to the formula.

$$\text{Digestible energy (Kcal/100g)} = \text{Protein\%} \times 4 + \text{Lipid\%} \times 9 + \text{Carbohydrate\%} \times 4$$

3.9.17 Amino acid and Fatty acid

Fatty acid analysis of the cultured shrimp whole body, feed and floc were done by the standard protocols (AOAC 1995), and amino acid analysis was done by using UPLC following standard protocols water corporation (2012).

3.9.18 Evaluation of EAA

The A/E ratio of each EAA (the concentration of each EAA as a percentage of the concentration of total EAAs) in floc, feed and shrimp samples were calculated according to the formula of Mente et al. (2002). The EAA score of each EAA was calculated as the A/E ratio of the floc or feed and the corresponding A/E ratio in the shrimp whole body. The EAA index (EAAI) was calculated as the geometric mean of the EAA scores using the established formula (Penaflorida 1989).

$$EAAI = (aa_1/AA_1 \times aa_2/AA_2 \times \dots\dots\dots \times aa_n/AA_n)^{1/n}$$

where aa is the A/E ratio of each EAA in the floc or feed, AA is the corresponding A/E ratio in the shrimp whole body and n is the number of essential AAs. To calculate EAAI, the values for the EAA score were constrained between 0.10 and1.00 (Penaflorida1989).

3.10 Enzyme analysis

3.10.1 Extraction of enzyme from floc

Enzyme extraction from floc sample was done by EDTA extraction method (Yu et al., 2007). The floc was settled for one and half hour at 4°C and supernatant was removed carefully from the sample. The sediment was then centrifuged at 2000 rpm 15 minutes. The collected bottom sediment was re-suspended to its original volume using a pH 7 buffer solution consisting of Na_3PO_4, NaH_2PO_4, NaCl and KCl. Again this sample was centrifuged at 5000 rpm @ 15 minutes with solid phase collection. Then floc was extracted using EDTA (2%, W/V) at 4ºC for 3 hours.

3.10.2 Lipase

The 'Lipase activity' assay was conducted by the method of Cherry and Crandel, (1932). Two test tubes were labelled as test and control, in each test tubes 3 ml of the distilled water and 1 ml of homogenate was added. One of the tube (marked as control) was placed in boiling water to inactive the lipase activity. Then 0.5 ml of buffer solution (Phosphate buffer pH 7) and 2 ml of olive oil emulsion was added in both tubes. The tubes were mixed and incubated at 37°C. After 24 hour's incubation 3 ml of 95% alcohol and 2 drop of phenolphthalein solution was added. Each test tubes were titrated with 0.05N NaOH until the appearance of permanent pink colour.

3.10.3 Amylase

The 'Amylase activity' was determined by Clark, (1964). A 20% of homogenate was prepared with sucrose solution. 1 g of sample was taken and homogenized with 0.25 M sucrose solution.

0.1 ml substrate (2% starch) and 1.8 ml of phosphate buffer (in blank 0.1 ml of distilled water) was taken along with 0.1ml of 20% homogenate for enzyme analysis. This mixed sample was incubated at 37°C for 30 minutes and 2 ml of DNS was added. To stop the enzyme activity, mixture was placed in boiling water and diluted with 10 ml distilled water for reading OD at 540 nm.

3.10.4 Protease

10 % of homogenate was prepared with phosphate buffer (pH 7.5 and 0.1 M). 0.5 ml of Tween 20 and 1 ml of homogenized sample was taken for the extraction of enzyme. The suspension was centrifuge completely to obtain it as clear enzyme solution. Then diluted with distilled water.

1 ml of final sample along with 1 ml of pre-incubated substrate was incubated in 37°C water bath for 20 minutes. After incubation 4ml of TCA was added to stop the reaction. Sample was placed in 50ºC water bath at 15 minutes to catalyze the reaction. The sample was centrifuged @ 3500rpm for 5 minutes, and 3 ml supernatant was mixed with 1 ml of 2N NaOH. OD reading was taken @ 440 nm after the color change in the sample.

$$U = (G \times F \times 1000) / t$$

Where, G : mg of Azocasein, F: dilution factor , t: reaction time

3.10.5 Cellulase

10 % of homogenate was prepared with phosphate buffer (pH 7.5 and 0.1 M). 0.5 ml of Tween 20 and 1 ml of homogenized sample was taken for the extraction of enzyme. The suspension was centrifuge completely to obtain it as clear enzyme solution. Then diluted with distilled water.

1 ml of final sample and 3 ml preheated substrate was taken in test tube and vortexed for 15 seconds. This mixture was incubated by placing test tubes in 50°C water bath at 20 minutes.2 ml of stop reagent and 1 ml of Ferricyanide was added in incubated sample. The mixture was vortexed for 15 seconds and placed in 100°C water bath to arrest the reaction enzyme. The sample was centrifuged at 4800 rpm @ 5 minutes to obtain the clear solution and OD was taken at 420 nm.

U= (4G x F) / t

Where, G : glucose concentration, F: dilution factor, t: reaction time

3.11 Floc parameters

Floc volume was measured using Imhoff Cone where the biofloc suspension was taken in the cone and allowed to settle for 20 minutes. The volume settled at the bottom was measured in ml L^{-1} as the floc volume (Avnimelech and Kochba 2009). Floc size and shape were recorded under digital microscope.

TS determined by a known volume of well-mixed unfiltered sample dried in hot air oven to a constant weight at 103 to 105°C (APHA, 2005) and TDS determined by well-mixed filtered sample through a weighed standard glass-fiber and dried in hot air oven to a constant weight at 103 to 105°C (APHA, 2005). TSS was analysed by a known volume of well-mixed sample filtered through a weighed standard glass-fibre filter. The residue retained on the filter was dried in a hot air oven to a constant weight at 103 to 105°C (APHA, 2005). The solid parameters such as TS, TDS and TSS was expressed in mg L-1

3.11.1 Specific surface area determination

The specific surface area of the fiocs was determined by dye adsorption technique. The dye adsorption experiments were carried out in 200 ml flasks. Each flask contained 100ml of mixture, consisting of 20 ml biofloc sample with dye stock solution (0.29mmoll^{-1}). The flasks were shaken vigorously for 90 minutes in a water bath at a constant temperature 20°C, to attain equilibrium condition. The contents of the flasks were centrifuged at 3000 rpm for 15 minutes. Supernatant was collected and diluted before spectrophotometric determination. The optical density reading was recorded at 505 nm.

S = Y X N X A

where

S = specific area (m^2 g^{-1})

Y = adsorbed dye (mol /g)

N =Avogadro's number (6.023×10^{23} molecule mol^{-1})

A = area covered by each dye molecule (m^2 per molecule). = 1960 μm

3.11.2 Dehydrogenase activity

1 ml of biofloc sample was pipetted in four test tubes. Tris buffer (2.5 mL) and TTC-glucose solution (1 ml) were added to the sample tubes (1 ml of distilled water incontrol tube). The pH was adjusted to 7 using 1.0 N HCl and test tubes were gently swirled to mix the content. The tubes were incubated in an environmentally

controlled incubator (model number 2020, VWR International, Cornelius, Oregon, USA) at 30°C for 1 hour. The tubes were removed and centrifuged (IEC CentraCL2, Thermo Electron Corporation, Mississauga, Ontario, Canada) for 10 minutes to separate the cells from other medium components. TF extraction was carried out three times using 2.5 mL of ethanol. All samples were vortexed (Thermolyne Maxi Mix, Thermolyne Corporation, Hampton, New Hampshire, USA) to disrupt cell walls and leach TF from cells followed by centrifugation to separate the cells at the bottom. Supernatants from the three extractions were combined and absorbance of the combined supernatants was measured at 484nm.

3.11.3 Floc Density Index

FDI is the ml of floc which occupies a volume of 100 ml after 30 minutes of settling.

$$FDI = [(\text{ml settled in 30 min}) \times 100) / \text{floc concentration}]$$

3.11.4 Floc concentration

Floc concentration is the volume (ml) difference between the filtered and -unfiltered mixed liquid suspended matter. The formula for FC is

Floc concentration = Volume of mixed liquid suspended matter – Filtered volume

3.11.5 Floc density

Floc density is the ratio between the floc concentration and floc volume.

$$\text{Floc density} = \text{floc concertation} / \text{floc volume}$$

3.11.6 Floc Volume Index

FVI is used by operators to determine and compare floc settlability.

$$FVI = (1/ \text{floc volume}) \times 100$$

3.11.7 Pore volume

Pore volume defined as difference between the bulk volume and solid volume.

$$\text{Pore volume} = \text{Floc volume} - \text{Floc concentration}$$

3.11.8 Porosity

The total porosity is defined by the ratio of the volume of void space to the total or bulk volume of the materials.

$$\text{Porosity} = (1 / \text{pore volume})$$

Porosity can be expressed as % of the bulk volume of the materials

3.11.9 Floc Settling velocity

A special jar was designed to measure the individual floc settling velocity (Megara et al., 1976) with a settling column connected at the bottom of the jar. The column was 50 cm in length and filled with mixed liquid. Settling flocs were photographed and their velocities were estimated photographically by the distance travelled and the corresponding time taken for the traveling. The floc size was obtained simultaneously in the same process.

3.11.10 Floc Pore Radius

Mean floc pore radius is the characteristics described the fine floc physical structure. Floc pore radius is measured by using the equations derived by Broekhoff et al. (1970),

$$r = (2.5 \text{ X specific surface area X } 10^6) / \text{pore volume}$$

3.12 Apparent digestibility

The apparent digestibility of biofloc meals was estimated by the indirect method using chromic oxide (Cr_2O_3) as the inert marker at 10 g/kg of the diet. Initially, the concentration of Cr_2O_3 in the finished diets and in shrimp feces was used to determine the apparent digestibility coefficient (ADC)

$$ADC = 100 - [\ 100\ (\%\ Cr_2O_{3\,f} / \%\ Cr_2O_{3\,d}) \times (\%\ N_f / \%N_d)]$$

in which

ADC = apparent digestibility coefficient of nutrient (crude protein and lipid).

Cr_2O_{3d} = concentration (in %) of chromic oxide in the diet

Cr_2O_{3f} = concentration (in %) of chromic oxide in shrimp feces

N_d = concentration (in %) of nutrient (crude protein and lipid) in the diet

N_f = concentration (in %) of nutrient (crude protein and lipid) in shrimp feces.

Subsequently, the digestibility of crude protein and lipid of each individual ingredient under evaluation was calculated (Cho et al., 1982; Bureau et al., 1999).

3.13 Economics

The study also documented the techno economic feasibility of the nursery rearing of shrimp in AMF systems. The fixed cost and the variable cost were calculated according to the input price in Tamil Nadu where justification behind the economics of the production quantity transparent to creditability of the culture system for establishing the confidence for adoption from the outcomes of the study.

CHAPTER – 4

Results

4.1 Water quality parameters

The experimental trial in nursery raceways was conducted for 30 days. Water quality parameters observed were temperature (ºC), dissolved oxygen (mg L^{-1}), pH, total alkalinity (mg L^{-1}), salinity (ppt), total hardness (mg L^{-1}), calcium (mg L^{-1}), magnesium (mg L^{-1}), Ammonia-Nitrogen (NH_3-N) (mg L^{-1}), Nitrite- Nitrogen (NO_2-N) (mg L^{-1}), Nitrate- Nitrogen (NO_3-N) (mg L^{-1}), Ortho phosphate (PO_4^{3+}) BOD and COD. The values obtained were shown in the table 4.1.

The water temperature of the different experimental groups at 6.00 hours,12.00 hours and 18.00 hours ranged from 25°C – 31°C during the experimental period. There was no significant difference ($P>0.05$) between the treatments for temperature. Due to heavy intensive aeration, dissolved oxygen level between the treatments were in the range of 4–6 ppm and it was not significant between the treatments. The salinity was maintained at 19–20 ppt throughout the whole experimental period and there was no significant difference between the treatments.

There was a mild fluctuation in pH of the culture water based on the dissolved oxygen and temperature of the system. It ranges between 7.5–8 in all the treatments and no significant difference was found between the treatments.

Total alkalinity level raised from 140 mg L^{-1} at the initial period of an experiment to 200 mg L^{-1} at the end of the trial as shown in figure 4.1 H. A significant difference was observed between the treatments and lower level of alkalinity was found in control and higher in other treatments.

Total hardness was recorded in the range of 1500 to 2500 mg L^{-1}. A significant difference was noticed in T1 compared to T2 and C. The calcium content of the water was in the range of 250 to 350 mg L^{-1} and magnesium content was 270 to 330 mg L^{-1} as represented in the fig 4.1 E, 4.1G, 4.1F. The calcium and magnesium content ratio was maintained in the range of 1.2 to 2.5 and there was no significant difference between the treatments.

The nitrogenous compound such as ammonia –nitrogen (NH_4^+) was in the range of 0.18 to 0.4 mg L^{-1}, Nitrate nitrogen concentration fluctuated from 3 to 5.5 mg L^{-1} and nitrite nitrogen concentration was varied between 5.2 to 7 mg L^{-1}. There was a significant difference between the treatments and control in all the three nitrogen forms (Fig 4.1 I, J, K). Control has highest mean value of Ammonia, Nitrate and Nitrite whereas T1 and T2 has lowest mean values of nitrogenous compounds.

BOD level in treatments were significantly different between treatments and control group. However, no significant difference was found between treatments and control groups in COD. Changes in BOD and COD level were shown in the fig 4.1Q. R.

Table 4.1 Water quality parameters of experimental groups for the 30 days of culture trial

Parameter	Control	*Insitu (T1)*	*Exsitu (T2)*
DO (ppm)	5.53 ± 1.48^a (3.2 – 5.6)	6.33 ± 3.61^a (5.2 –8.6)	6.58 ± 3.32^a (5.2 – 7)
CO_2 (ppm)	8.89 ± 0.66^b (0–10)	5.27 ± 1.06^a (0-8)	4.17 ± 0.72^a (0–9)
Temperature (°C)	27.84 ± 1.34^a (25–30)	27.03 ± 1.05^a (25–30)	27.05 ± 1.14^a (25–30)
Salinity (ppt)	19.41 ± 1.56^a (19.5–21.5)	19.63 ± 0.48^a (19.5–21.5)	19.96 ± 0.29^a (19.5–21.5)
pH	7.63 ± 0.30^a (7–8.5)	7.67 ± 0.25^a (7–8.5)	7.69 ± 0.23^a (7–8.5)
Alkalinity (mg/L^{-1})	115.71 ± 33.79^a (46–180)	162.55 ± 19.45^b (124–200)	163.13 ± 23.18^b (139–230)
Hardness (mg/L^{-1})	2073.55 ± 888.07^b (220–3500)	1637.74 ± 809.23^a (200–3200)	2223.87 ± 625.6^b (160–3200)
Calcium (mg/L^{-1})	332.81 ± 136.01^a (148–735)	271.5 ± 93.87^a (126–486)	277.88 ± 86.88^a (147–484)
Magnesium (mg/L^{-1})	331.26 ± 161.68^a (53–623)	301.08 ± 153.14^a (115–618)	383.06 ± 121.29^a (186–647)
Ammonia ($N\text{-}NH_3$) (mg / L^{-1})	0.14 ± 0.02^b (0.017–1.169)	0.07 ± 0.01^a (0.001–0.7)	0.09 ± 0.02^a (0.001–0.8)
Nitrite ($N\text{-}NO_2$) (mg / L^{-1})	6.74 ± 1.37^b (0.1–11)	5.98 ± 1.72^a (0.1–12)	5.28 ± 1.52^a (0.1–12)
Nitrate ($N\text{-}NO_3$) (mg / L^{-1})	5.18 ± 1.72^b (0.6–13)	5.09 ± 1.11^a (0.3–10)	5.23 ± 1.24^a (0.3–10)
Phosphate (PO_4^{3+}) (mg /l)	12.104 ± 2.791^b (4–15.6)	7.222 ± 2.64^a (5–10.9)	7.821 ± 2.571^a (5–11)
BOD_5	2.64 ± 1.04^a (1.7–5.28)	5.37 ± 2.21^b (0.5–8)	6.68 ± 2.78^b (3.3–11.2)
COD	8.75 ± 0.76^a (5–12)	10.25 ± 3.22^a (3.7–25.58)	8.26 ± 2.27^a (2.8–18.97)

4.2 Microbial activity in culture water

Microbial load observed in the culture water during the experimental period were showed in table 4.2 and figure 4.2. There was a significant difference in the AMF culture water and control. Total Heterotrophic bacterial count was lower in the control compared to T1 and T2.THB count in T2 was 136.5 ± 0.04 x 10^4 CFU/ml while in control it was only 19.5 ± 0.01 x 10^4 CFU/ml. Total Vibrio count was significantly different between control and treatments. The level of Vibrio count in T1 was found to be higher to the log value of 5.32 ± 0.045 x 10^4 CFU/ml and lower value was observed in 1.56 ± 0.01 x 10^4 CFU/ml.

Table 4.2 THB and TVC of the culture water the experimental tanks

DOC	Total Heterotrophic Bacteria (CFUX 10^4/ml)			Total Vibrio Count (CFU X 10^4/ml)		
	Control	***Insitu (T1)***	***Exsitu (T2)***	**Control**	***Insitu (T1)***	***Exsitu (T2)***
6	0.4±0.01 [a]	0.93±0.01 [b]	0.82±0.01 [b]	0.03 ± 0.001 [a]	0.18± 0.001 [b]	0.13 ±0.001 [b]
12	0.52±0.05 [a]	1.1±0.04 [b]	1.63±0.012 [b]	0.05 ±0.001 [a]	0.86 ±0.01 [b]	0.62 ±0.05 [b]
18	1.2±0.01 [a]	3.3±0.01 [b]	7.1±0.018 [b]	0.12 ± 0.05 [a]	1.69 ±0.01 [b]	1.01 ±0.01 [b]
24	6.8±0.06 [a]	10.3±0.015 [b]	13.83±0.05 [b]	0.85 ±0.04 [a]	3.8 ±0.025 [b]	3.31 ±0.05 [b]
30	19.5±0.01 [a]	92.5±0.12 [b]	136.5±0.04 [b]	1.56 ±0.01 [a]	5.6 ±0.054 [b]	5.32 ±0.045 [b]

Values (Mean ± SE) in the same row with different superscript differ significantly (Duncan Multiple Range Test ($p<0.05$)). All the values in percentage were Arsine transformed for ANOVA

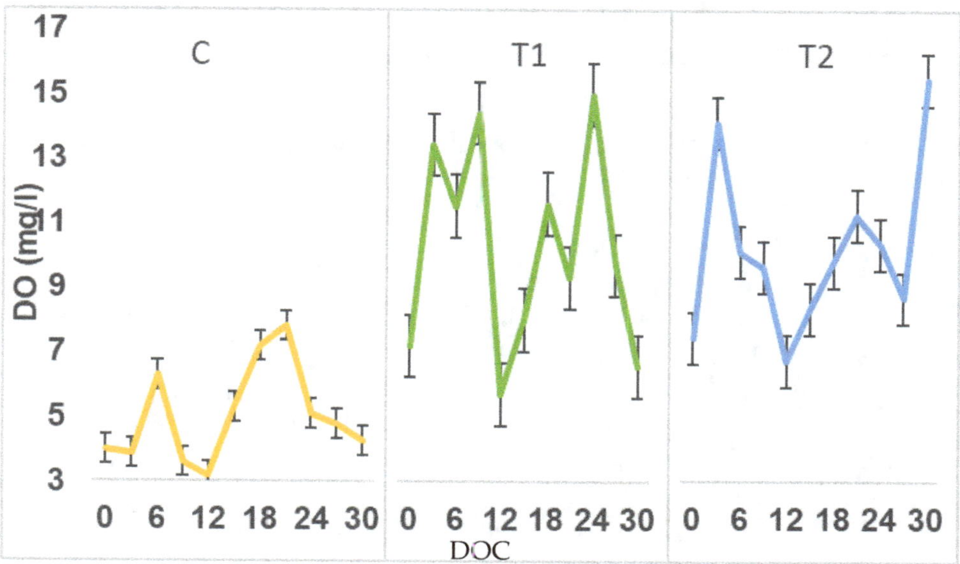

A

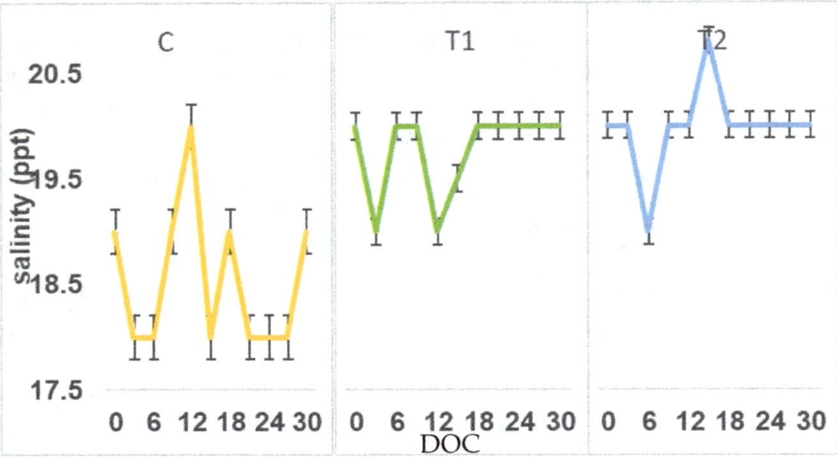
C
T1
T2
salinity (ppt)
20.5
19.5
18.5
17.5
0 6 12 18 24 30
0 6 12 18 24 30
0 6 12 18 24 30
DOC

B

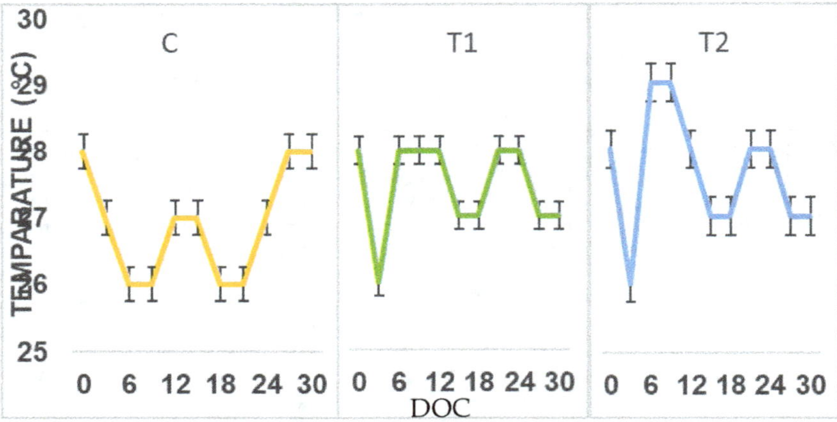
C
T1
T2
TEMPARATURE (ºC)
30
29
28
27
26
25
0 6 12 18 24 30
0 6 12 18 24 30
0 6 12 18 24 30
DOC

C

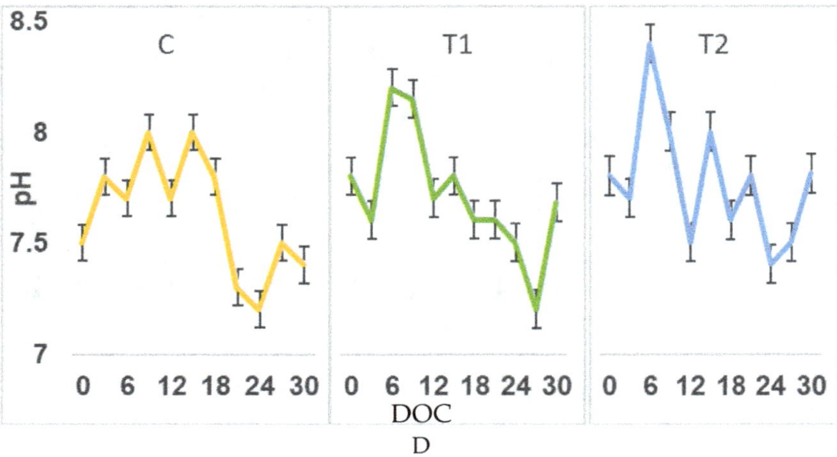
C
T1
T2
pH
8.5
8
7.5
7
0 6 12 18 24 30
0 6 12 18 24 30
0 6 12 18 24 30
DOC

D

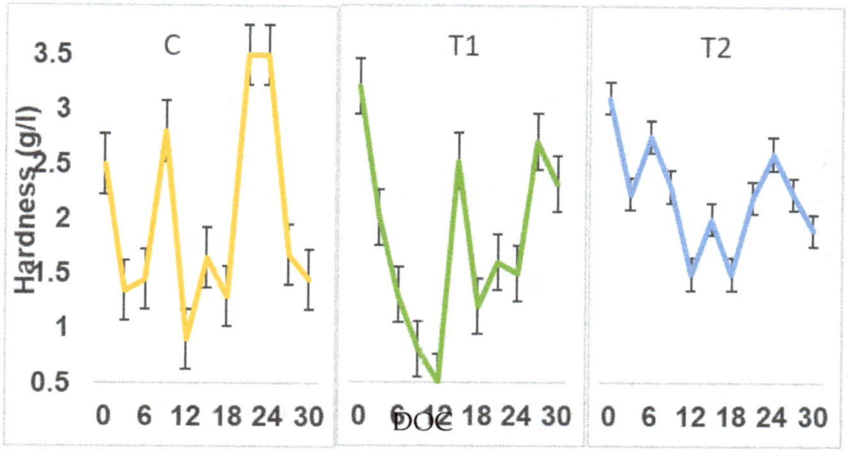
C
T1
T2
Hardness (g/l)
3.5
3
2.5
2
1.5
1
0.5
0 6 12 18 24 30
0 6 12 18 24 30
0 6 12 18 24 30
DOC

E

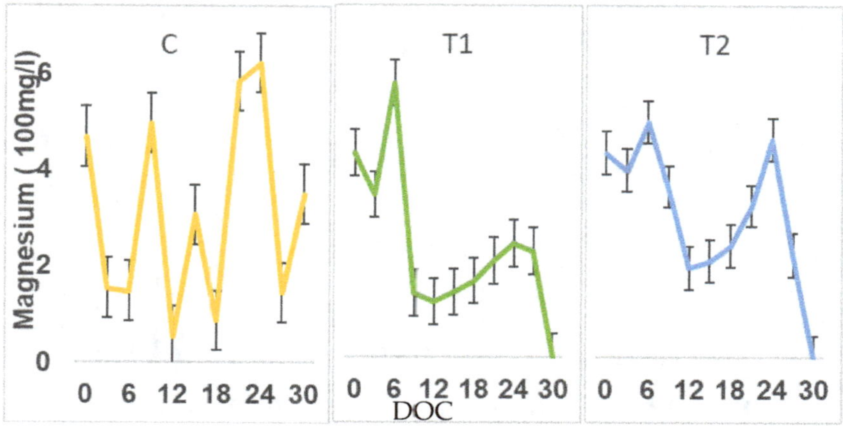
C
T1
T2
Magnesium (100mg/l)
6
4
2
0
0 6 12 18 24 30
0 6 12 18 24 30
0 6 12 18 24 30
DOC

F

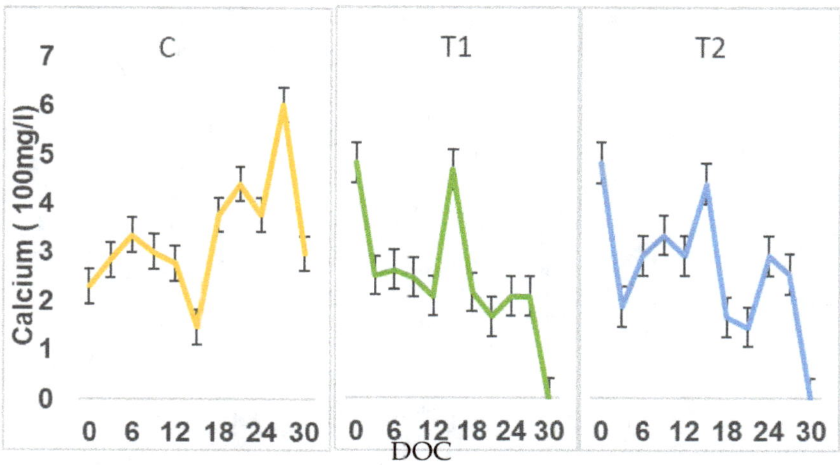
C
T1
T2
Calcium (100mg/l)
7
6
5
4
3
2
1
0
0 6 12 18 24 30
0 6 12 18 24 30
0 6 12 18 24 30
DOC

G

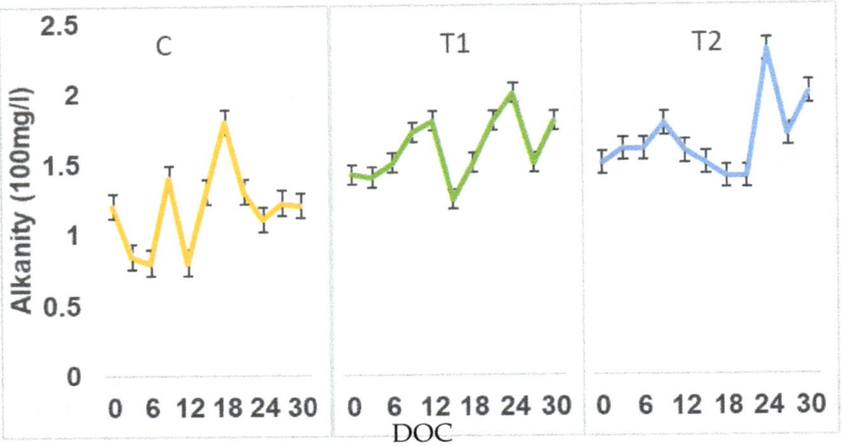
C
T1
T2
Alkanity (100mg/l)
2.5
2
1.5
1
0.5
0
0 6 12 18 24 30
DOC

H

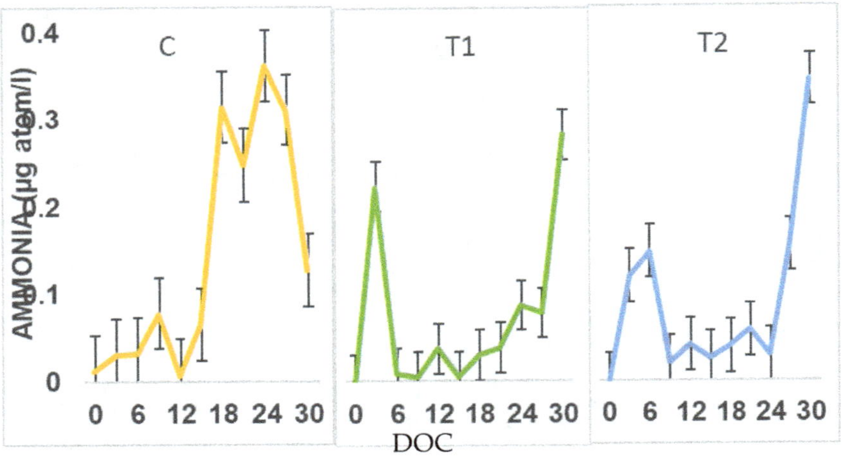
C
T1
T2
AMMONIA (µg atom/l)
0.4
0.3
0.2
0.1
0
0 6 12 18 24 30
DOC

I

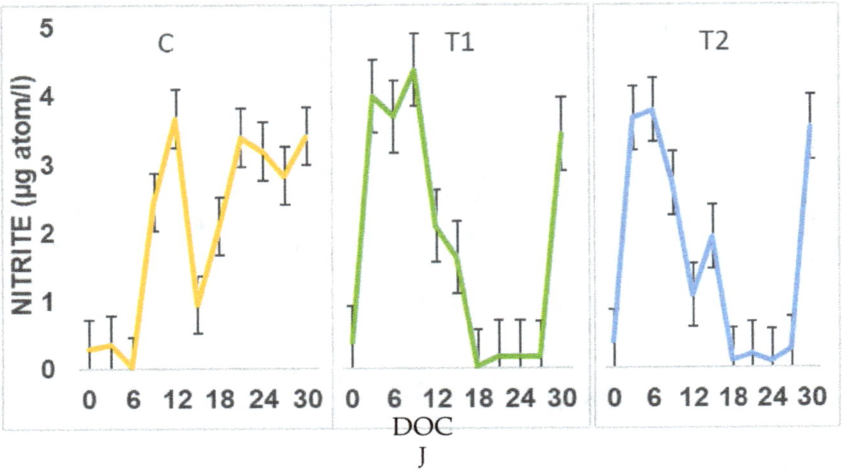
C
T1
T2
NITRITE (µg atom/l)
5
4
3
2
1
0
0 6 12 18 24 30
DOC

J

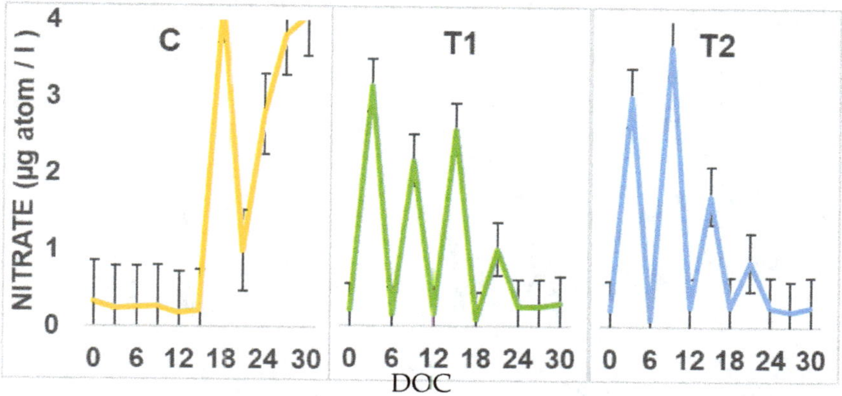
NITRATE (µg atom / l)
C
T1
T2
0 1 2 3 4
0 6 12 18 24 30
DOC

K

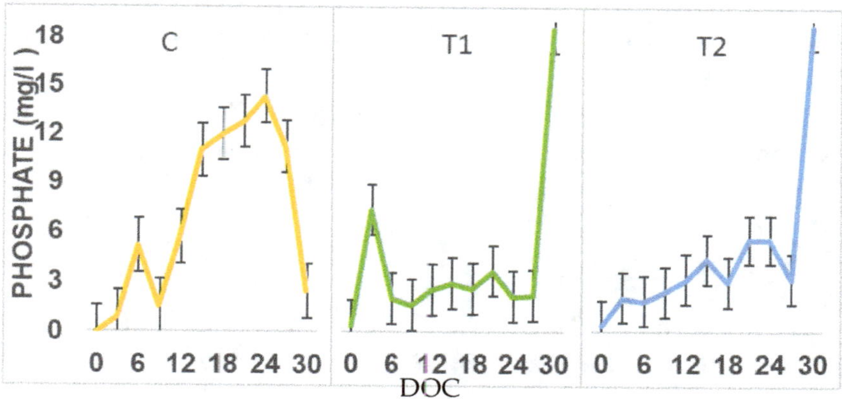
PHOSPHATE (mg/l)
C
T1
T2
0 3 6 9 12 15 18
0 6 12 18 24 30
DOC

L

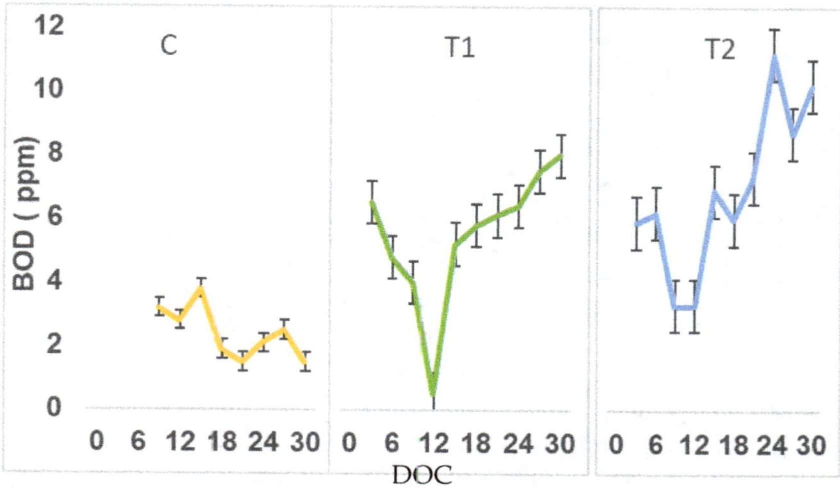
BOD (ppm)
C
T1
T2
0 2 4 6 8 10 12
0 6 12 18 24 30
DOC

Q

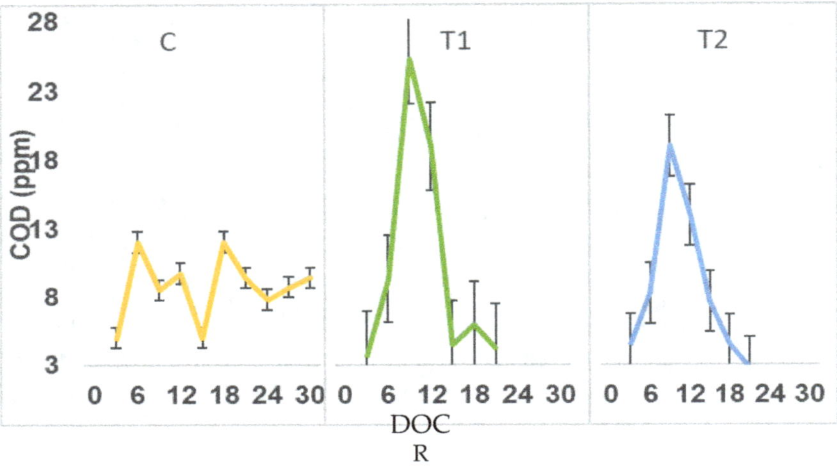

Fig. 4.1: Water quality parameters of experimental groups for the 30 days of culture trial

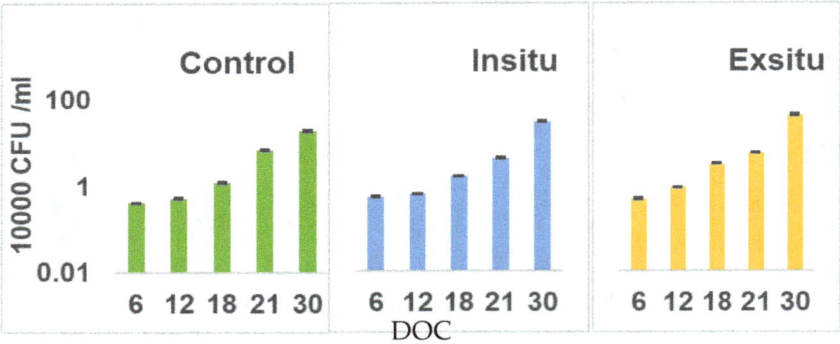

Fig. 4.2A: Total Heterotrophic Bacteria

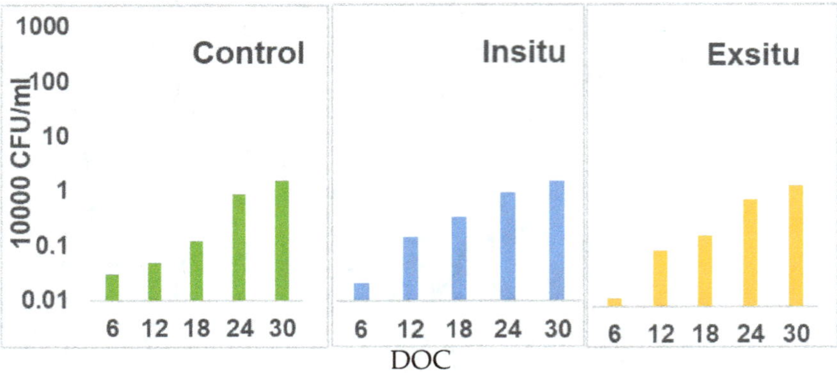

Fig. 4.2B: Total Vibrio Count

4.3 Growth parameters

The mean body weight of *P.vannamei* at different treatment groups were represented in Table.4.3. The result indicates both T1 and T2 system showed significantly higher growth rate than control ($P < 0.05$). The highest growth in terms of body weight was recorded in T2 system followed by T1 and control. The initial body weight was in the range from 0.02 ± 0.01 g while final mean body weight varied from 1 g to 1.035 g in different treatment groups as shown in the fig 4.3.

The survival rate of the animals in the different treatment groups was represented in Fig 4.2.2. All the AMF based treatment groups obtained similar ($P<0.05$) survival compared to control. The total initial biomass recorded in each treatment groups were similar (0.02± 0.01) at the beginning of the culture period. The final total biomass was maximum in the T1 system followed by T2 and control. The percentage biomass was significantly($P<0.05$) different between control and treatments.

The value of FCR, FER SGR were significantly different between control and other treatments. Growth rate per day range from a maximum of 33.84 mg in T2 system with SGR 13.15% and a minimum of 32.8 mg in control with SGR 13.05% was represented in fig 4.2.2 and table 4.3. FCR, FER and PER of all the treatment groups are represented in Table 4.3 and Fig 4.2.2. FCR was minimum in T2 (1.044) and maximum in control (1.076). FER was maximum in T2 (0.957) and minimum in control (0.929).

PER was higher in T2 (0.0317) and C (0.02813) compared to T1 (0.02389). Animal fed with AMF based feed (T2) shown significantly higher SGR, FER and better FCR compared to other treatments.

Table. 4.3 Zoo technical performance of *P.vannamei*.

S. No.	Growth parameters	Treatments		
		control	T1	T2
1.	Initial body weight (g)	0.02 ±0.003 ^{a}	0.02 ±0.001^{a}	0.02 ±0.002 ^{a}
2.	Final body weight (g)	1.0048 ^{b}±0.005	1.0322 ^{a}±0.005	1.0352 ^{a}±0.005
3.	Final length (cm)	5.172^{b}± 0.028	5.344 ^{a}±0.058	5.32 ^{a}± 0.023
4.	Weight gain (g/ week)	11.489 ^{b}±0.043	11.809 ^{a}±0.023	11.844 ^{a}±0.017
5.	Percentage weight gain (%)	49.24^{b}±0.01	50.61 ^{a}±0.05	50.76 ^{a}±0.05
6.	Net weight gain (g)	0.9843 ^{b}	1.0122 ^{a}	1.0152 ^{a}
7.	Survival rate (%)	79.99 ^{b}±0.14	88.88 ^{a}±0.57	89.44 ^{a}±0.57
8.	Total initial biomass (Kg)	0.9 ^{a}	0.9 ^{a}	0.9 ^{a}
9.	Total final biomass (Kg)	36.16 ^{b}±0.0	41.28 ^{a}±0.0	41.66 ^{a} ± 0.0
10.	% biomass (%)	97.51 ^{a}	97.81 ^{a}	97.83 ^{a}
11.	Growth rate (mg/day)	32.82 ^{b}±0.011	33.74 ^{a}±0.057	33.84 ^{a}±0.057
12.	SGR (% day^{-1})	13.05 ^{b}±0.01	13.14 ^{a}±0.01	13.15 ^{a}±0.01
13.	FCR	1.076 ^{b}±0.004	1.047 ^{a}±0.006	1.044 ^{a}±0.006
14.	FER	0.9290 ^{b}±0.005	0.9544 ^{a}±0.18	0.9571 ^{a}±0.12
15.	PER	0.02813 ^{b} ±0.0	0.02389 ^{c} ±0.0	0.03172 ^{a}±0.0

Values (Mean ± SE) in the same row with different superscript differ significantly (Duncans Multiple Range Test ($p<0.05$)). All the values in percentage were Arsine transformed for ANOVA. *SGR: Specific Growth Rate; FCR: Feed Conversion Ratio; FER: Feed Efficiency Ratio; PER: Protein Efficiency Ratio.

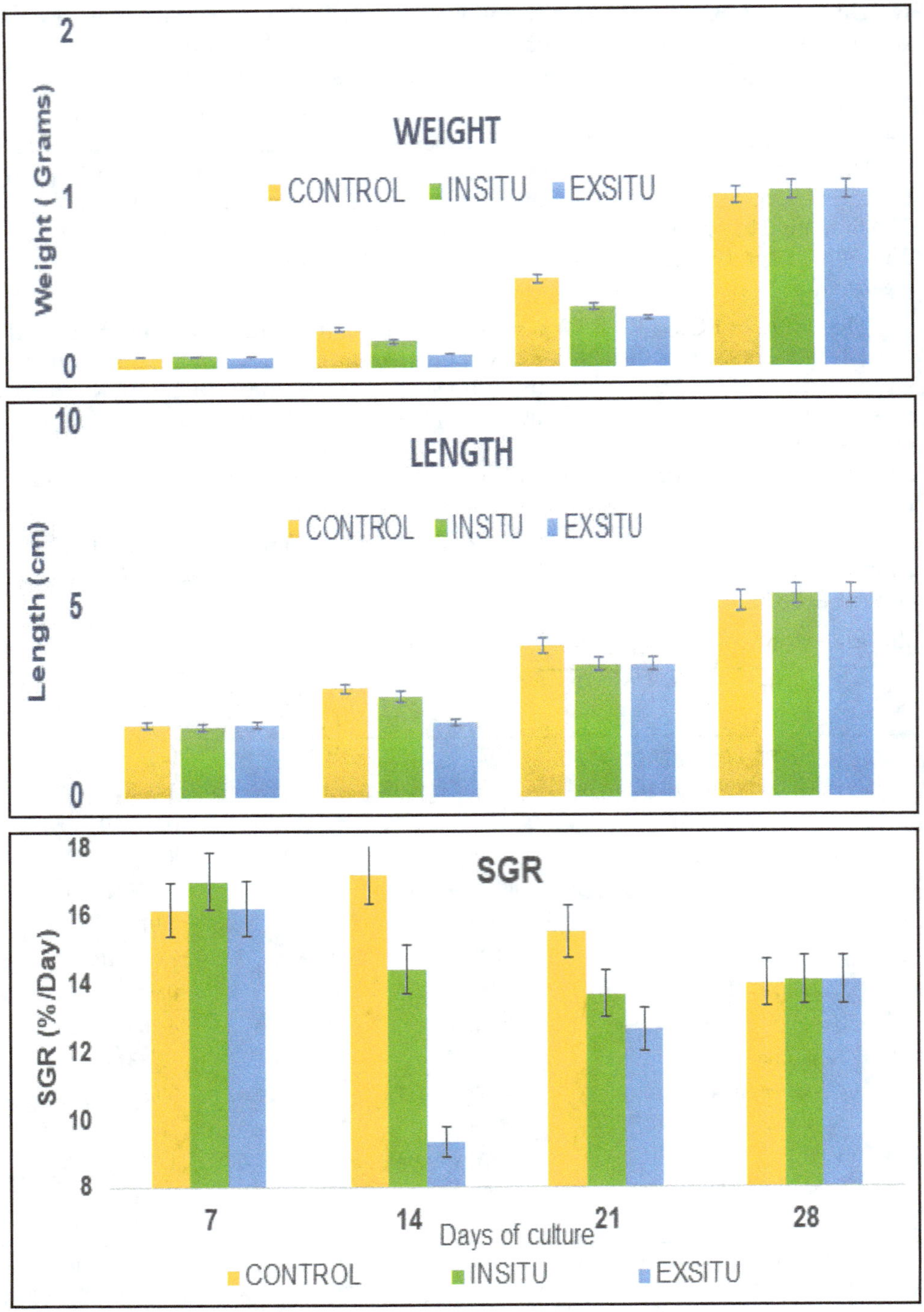

Fig. 4.3: Zoo technical performance of P.vannamei.

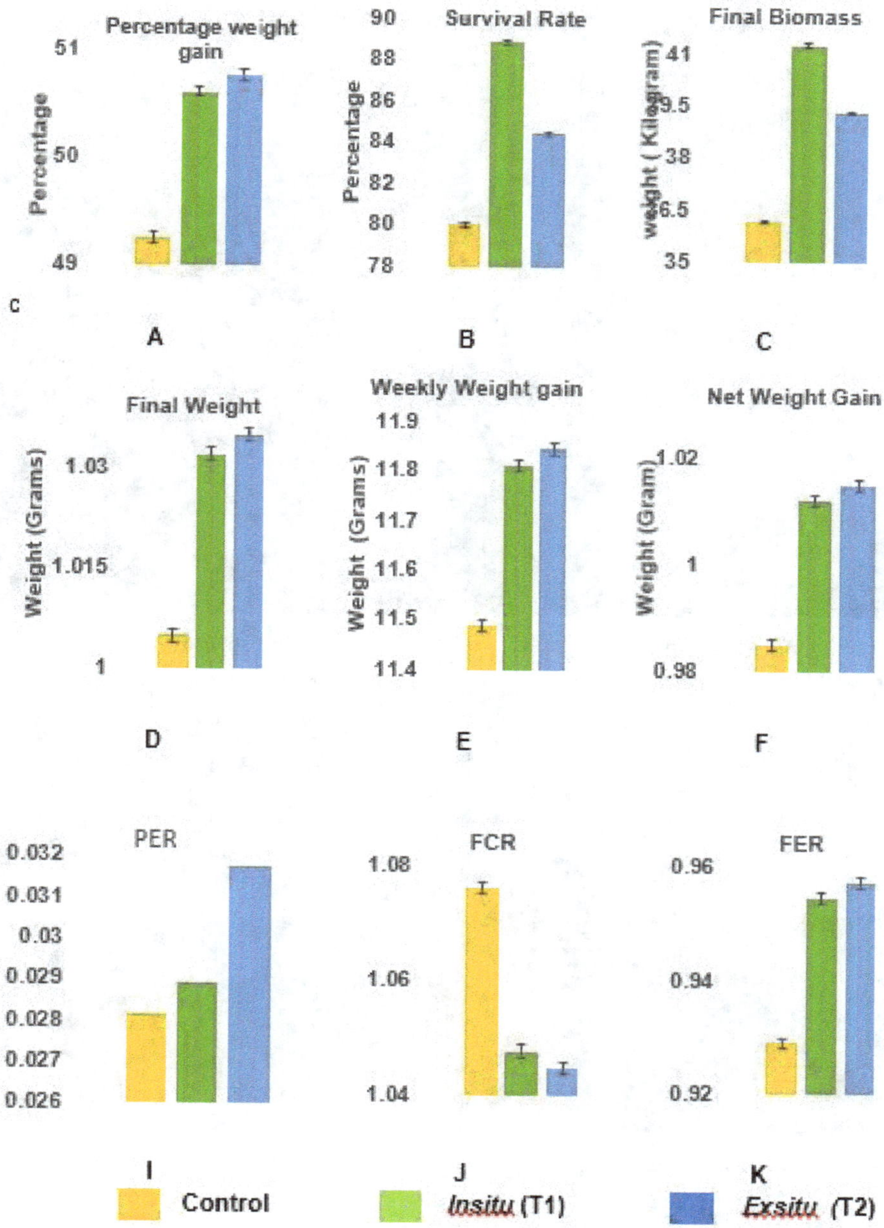

Fig. 4.2.2 : Zoo technical performance of P.vannamei.

Plate – 3 Experimental animal

3A – Shrimp PL 15

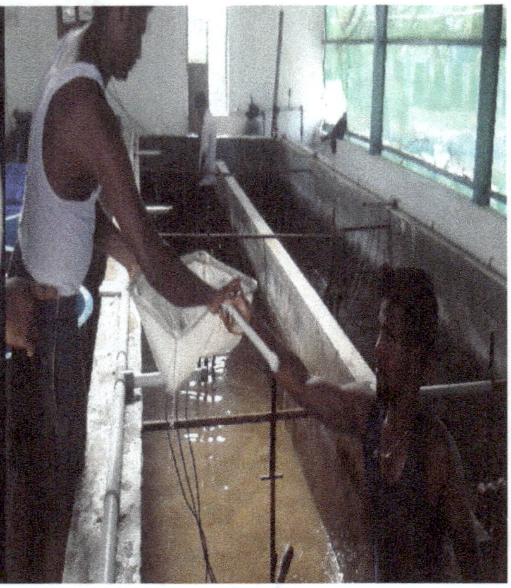

3B – Harvesting shrimps from raceway

3C – Initial weight

3D – Final weight

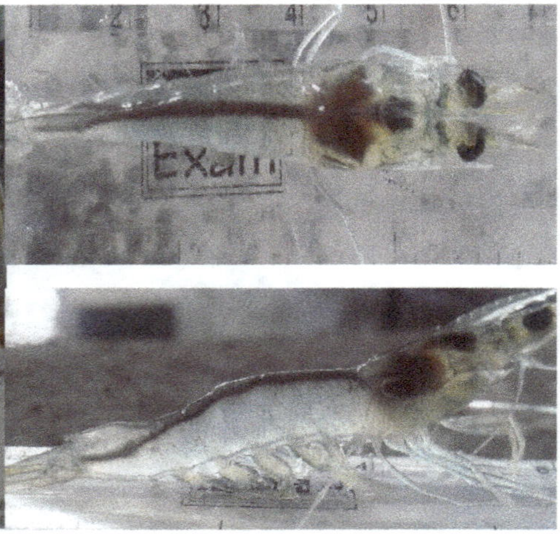

3E –Shrimp at DOC 30

Plate – 4 Experimental unit

4A – Control tank

4B – *Insitu* tank

4C – *Exsitu* tank

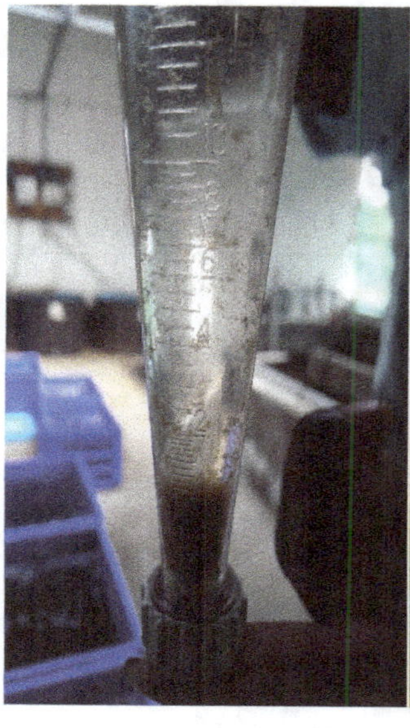

4D – Floc volume in Control

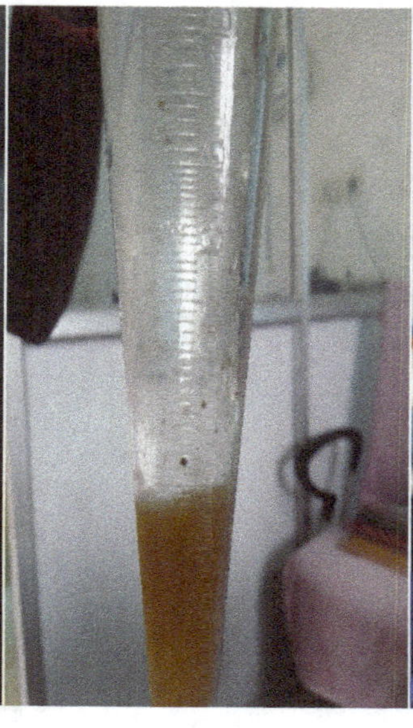

4E - Floc volume in *Insitu* (T1)

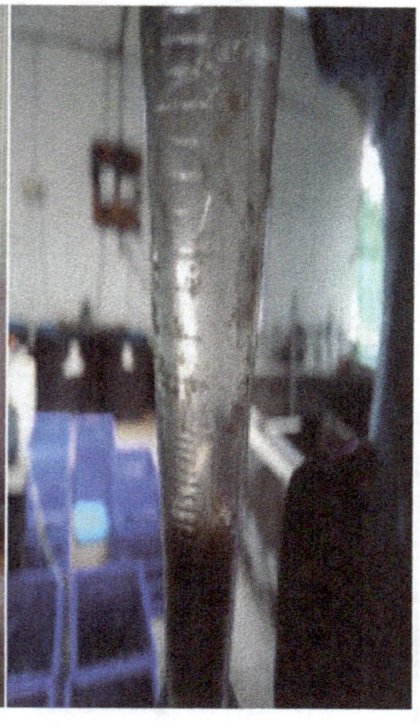

4F - Floc volume in *Exsitu* (T2)

4.3 RNA – DNA Ratio

RNA and DNA levels observed in muscle during the experimental period were given in Table 4.3. There was no clear trend for DNA and RNA content in the muscle of *P. vannamei* during the culture period in different treatment groups. In the first 15 days, DNA content was higher in all treatments however; it was reduced at the end of the culture trial in all the treatment groups except T1. At the end of the experiment lowest DNA content in the muscle was observed in control (4.7763 ± 0.0503) which was significantly lower than other treatments. RNA content of muscle tissue was found at decreased level in C and T2 throughout the trial. Meanwhile a gradual increase in RNA content was observed in T1 at the end of the experiment.

RNA - DNA ratio observed in different experimental groups during the culture period was shown in Table 4.3 and Fig 4.3. The ratio between RNA and DNA ranged from 0.20 to 0.32 in all the experimental groups. In the initial phase of the experiment the ratio was higher in all the treatment groups, then it reduced drastically towards the end of the experiment. RNA -DNA ratio showed a declining pattern at 15th and 30th day for all the experimental group's excepted T1. At the end of the experiment (30th day) control reared shrimps showed highest RNA -DNA ratio which was significantly different ($P<0.05$) from other treatment groups.

Table 4.4. RNA : DNA Ratio of *P.vannamei* reared in different treatments

DOC	15			30		
Treatment	C	T1	T2	C	T1	T2
DNA(µg/ ml)	4.3552 ± 0.0503^{b}	3.934 ± 0.1315^{a}	5 ± 0.1136^{c}	4.7763 ± 0.0503^{a}	5.4605 ± 0.1384^{c}	5.0789 ± 0.0679^{b}
RNA µg/ ml)	1.1412 ± 0.0107^{b}	0.7918 ± 0.0128^{a}	1.6077 ± 0.0131^{c}	1.0918 ± 0.0104^{b}	1.1014 ± 0.0187^{b}	1.0234 ± 0.0145^{a}
RNA:DNA	0.262 ± 0.005^{b}	0.201 ± 0.009^{a}	0.321 ± 0.006^{c}	0.2286 ± 0.0040^{b}	0.2018 ± 0.0084^{a}	0.2015 ± 0.0054^{a}

Values in the same row with different superscripts differ significantly ($P<0.05$) for each parameter. One way ANOVA was used following Duncan multiple range test in SPSS

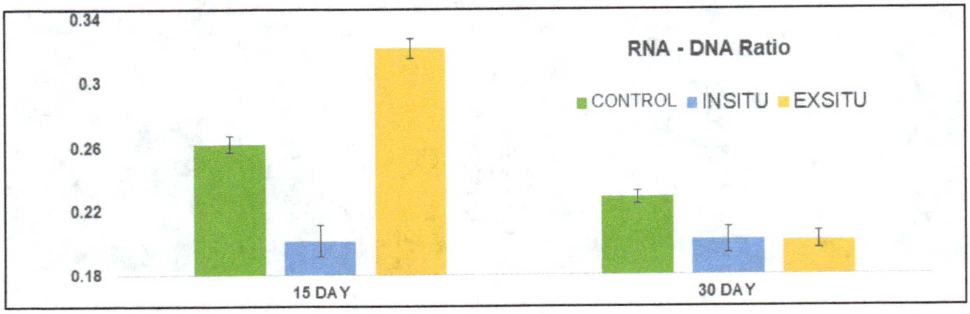

Fig. 4.5: RNA DNA ratio of P.vannamei reared in different treatments

4.4 Extra cellular enzyme activity

Extracellular enzymes including protease, amylase, lipase and cellulase were found in the biofloc collected from the treatments as showed in Fig 4.5.1. Protease, amylase, cellulase and lipase showed a significant difference (P<0.01) in all treatments. While higher enzyme activity was recorded in *Exsitu* (T2) treatment and lower activity was recorded in control as given in Table 4.5.1 and fig 4.5.1.

Table – 4.5. Extracellular enzyme activity (Unit/ µg protein/ minute) of the AMF produced in different AMF groups stocked with *P.vannamei*

Enzymes	C	T1	T2	P value
Protease	3.5498 ± 0.3470[a]	52.761 ± 0.5622[b]	49.697 ± 2.550[c]	**
Lipase	263.608 ± 9.3199[a]	481.7535 ± 98.337[b]	726.392 ± 104.6124[c]	**
Amylase	5.8110 ± 0.0616[a]	38.619 ± 0.5959[b]	43.3936 ± 1.2298[c]	**
Cellulase	-0.004047 ± 0.00035[c]	-0.3315 ± 0.00178[b]	-0.48750 ± 0.001721[a]	**

Values in the same row with different superscripts differ significantly (P<0.01) for each parameter. One way ANOVA was used following Duncan multiple range test in SPSS

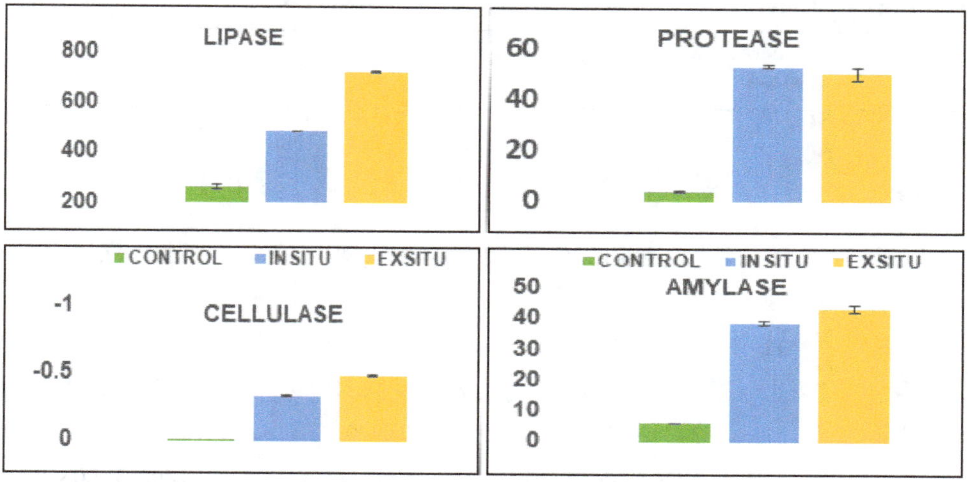

Fig 4.6 Extracellular enzyme activity of the AMF produced in different AMF groups stocked with P.vannamei

4.5 Digestive enzyme activity in shrimp visceral mass

Different digestive enzyme activities such as protease, amylase lipase and cellulase were observed in shrimp visceral mass as shown in table 4.4.1 and figure 4.4.1.

The specific activity of the protease in hepatopancreas stomach and intestine showed a significant difference among the treatments. Higher specific activity of protease was found in ex*situ* treatment and lower in Control.

The specific activity of amylase in hepatopancreas, stomach and intestine showed a significant difference in all treatments. Specific activity of amylase was higher in *exsitu* treatment and lower in control.

The specific activity of lipase in hepatopancreas, stomach and intestine showed a significant difference in all treatments. Lipase specific activity in hepatopancrease, stomach and intestine was observed higher in *exsitu* treatment and lower in control.

Cellulase specific activity in hepatopancreas, stomach and intestine showed a significant difference in all treatments. Hepatopancreas and stomach showed the higher activity of lipase in T2 treatment and control showed higher cellulase activity in the intestine.

4.5.1 Digestive enzyme activity between Treatments and Visceral organs

There was no significant difference in protease specific activity between the treatments and visceral parts. There was a significant difference in the specific activity of lipase, amylase and cellulase at $p < 0.05$ level and $p < 0.01$ level between the treatments and shrimp visceral organs.

Table 4.6 Specific activity of digestive enzymes (Units/ μg protein/minute) in the digestive organ of P.vannamei in different experimental groups at the end of the feeding experiments

Enzyme	Visceral mass	C	T1	T2
Protease	H	186.67±0.579[a]	1233.95±1.77[b]	2895.189±57.057[c]
	I	136.318±1.538[a]	1195.453±1.334[b]	2976.294±9.64[c]
	S	214.327±0.875[a]	1735.33±5.766[b]	3526.292±8.763[c]
Amylase	H	9.308±0.205[a]	144.425±3.313[b]	309.985±2.14[c]
	I	52.072±1.631[a]	285.895±1.024[b]	598.938±11.981[c]
	S	38.74±0.411[a]	196.862±2.709[b]	481.489±2.058[c]
Lipase	H	1729.933±105.161[a]	2125.383±144.614[b]	5620.892±353.034[c]
	I	1977.066±142.365[a]	1779.204±524.141[b]	7263.923±2795.885[c]
	S	878.696±35.873[a]	3257.311±186.989[a]	5508.475±55.03[b]
Cellulase	H	0.018±0[b]	0.009±0.001[a]	0.11±0.004[c]
	I	-0.009±0.004[a]	-0.044±0.018[b]	-2.197±0.026[c]
	S	0.131±0.001[a]	0.525±0.002[b]	0.949±0[c]

(H- Hepatopancreas, I- intestine, S- Stomach)

Values in the same row with different superscripts differ significantly ($P<0.05$) for each parameter.
One way ANOVA was used following Duncan multiple range test in SPSS

Table. 4.7 Comparison of specific activity of enzymes among three treatment and digestive organ on two way repeated measures ANOVA.

Enzyme activity	Treatment			Part			Interaction Treatment* Organ
	C	T1	T2	H	S	I	
Protease	1.791 [a]	1.041[a]	2.966 [a]	1.04011 [a]	1.825[a]	1.436[a]	NS
Amylase	33.373 [a]	209 [b]	463 [c]	1.54 [a]	2.39 [b]	3.12 [c]	**
Lipase	1.52 [a]	2.387 [a]	6.13 [b]	3.158 [a]	3.214[a]	3.673 [a]	*
Cellulase	0.047 [b]	0.163 [c]	-0.376[a]	0.045 [b]	0.534 [c]	-0.7498 [a]	**

(H- Hepatopancreas, I- intestine, S- Stomach) * P< 0.05 ** P<0.01 NS – Not significant Units are in seconds. The mean value followed by the different superscripted letter within factor indicates significant difference at (a<b<c). if the effects were significant. ANOVA was followed by turkey test. *P<0.05, ** P<0.01, NS- not significant

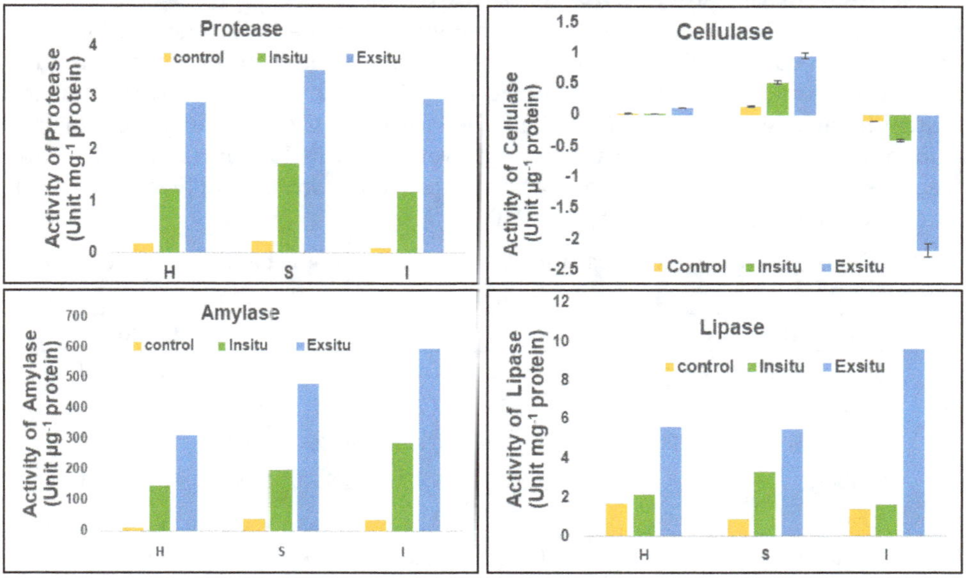

Fig. 4.7 specific activity of digestive enzymes in digestive organ of P.vannamei in different experimental groups

4.5.2 Biochemical composition

The proximate composition of shrimp whole body tissue, experimental diets, and AMF developed in the system were estimated by following the standard protocols (AOAC, 1995). The results were presented in the table. 4.6.1 and Fig.4.6.1

4.5.2 Biochemical composition of AMF

The crude protein in T2 AMF was relatively higher than T1. The fibre and total ash percentage in T2 floc was lower than the T1 floc. There were no striking changes in the composition of fat and moisture of T1 and T2 floc. The relative contents of Palmitic acid, Oleic acid, Linoleic acid, Linolenic acid and Arachidic acid was higher in T2 compared to T1. However C20:5n3 (Eicosapentaenoic acid) and C22:6n3 (Docosahexaenoic acid) were higher in T1. Lower level of n-3 fatty acid was found in the both treatments. The results were presented in the Table-4.8, 4.9, 4.10.

4.5.3 Biochemical composition of shrimp

The protein composition of shrimp whole body was found lower in T2 compared to other treatments. Moisture, fiber, fat and total ash composition did not show variant difference between the treatments. The relative contents of Palmitic acid, Oleic acid, Linoleic acid, Linolenic acid and Arachidic acid was higher in control followed by T2 compared to T1. However, C20:5n3 (Eicosa Pentaenoic Acid) and C22:6n3 (Docosa Hexaenoic Acid) was higher in T2 compared to control. The results were presented in the Table-4.8, 4.9, 4.10.. Similar level of n-3 and n-6 fatty acids were also recorded in all treatments ranging from 17-20 %.

4.5.4 Biochemical composition of shrimp feed

In shrimp diet the crude protein level remains the same in T1 and T2. Fiber, fat and total ash content didn't show variant difference between the treatments. The relative contents of Palmitic acid,Oleic acid, Linoleic acid, Linolenic acid and Arachidic acid was higher in T2 than T1. However, C20:5n3 (Eicosapentaenoic acid) and C22:6n3 (Docosahexaenoic acid) was higher in T1 compared to T2. AMF feed has lower level of n-3/n-6 fatty acid compared to commercial feed. The results were presented in the Table-4.8, 4.9, 4.10.

Table 4.8 The proximate composition of the (% dry weight basis) of the AMF, feed and shrimp

S. No	Parameters	Feed		Shrimp			Biofloc	
		Commercial	Biofloc	C	T1	T2	T1	T2
1.	Moisture	11.66	13.57	2.77	6.24	2.91	10.4	15.74
2.	Protein	36.354	35	65.47	63.27	42.78	27.93	27.93
3.	Fiber	2.85	3.62	5.43	4.66	5.08	3.77	1.69
4.	Fat	5.62	6.77	2.91	4.26	2.79	0.88	1.13
5.	Total ash	8.99	10.94	16.6	12.68	13.13	38.64	30.86
6.	GE*	4083	3885	4464	4519	4216	2547	2658

GE- Gross Energy

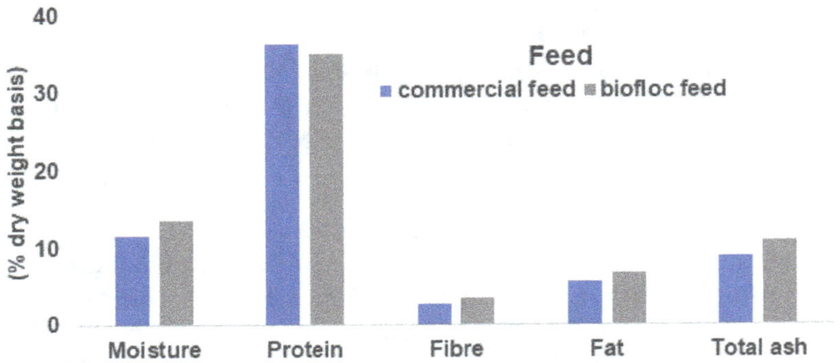

Fig. 4.8 A - Proximate composition of the experimental diets

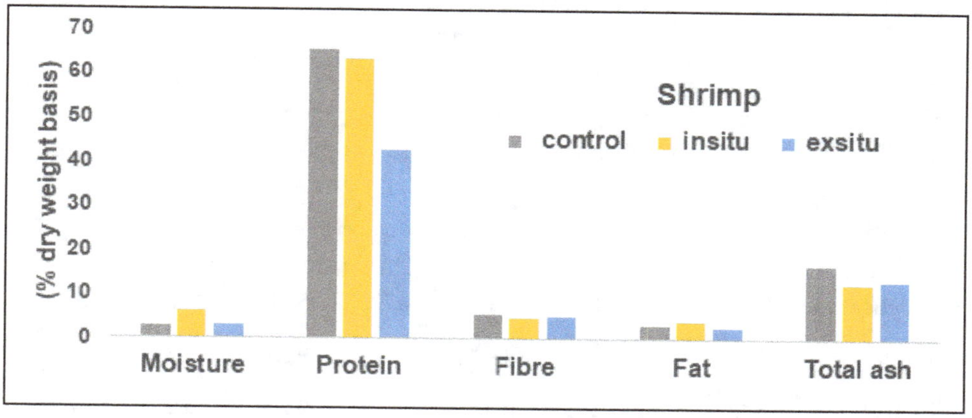

Fig. 4.8 B Proximate composition of the cultured animal in different treatments

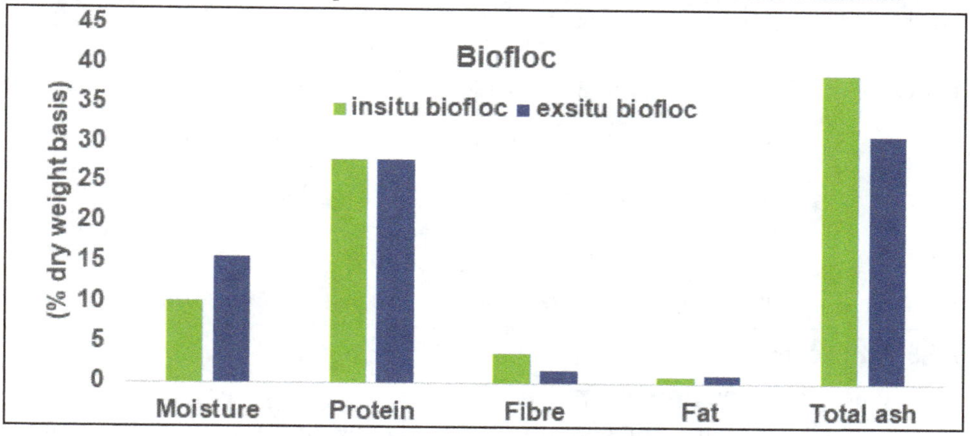

Figure 4.8 C Proximate composition of the AMF produced in the treatments

Table 4.9 Fatty acid profile of the of the AMF, feed and shrimp. (% dry weight basis)

S. No	Parameters	Feed		Shrimp			Biofloc	
		Commercial	Biofloc	C	T1	T2	T1	T2
1.	Myristic acid	3.6	2.87	1.02	1.01	0.85	7.33	2.29
2.	Palmitic acid	22.29	19.87	20.71	21.24	19.97	22.91	25.66
3.	Stearic acid	5.75	4.14	11.6	10.91	11.48	8.76	8.97
4.	Olelic acid	18.45	19.01	21.2	20.91	21.37	30.93	28.65
5.	Linoleic acid	25.77	35.89	17.36	18.2	17.38	9.04	15.47

S. No	Parameters	Feed		Shrimp			Biofloc	
		Commercial	Biofloc	C	T1	T2	T1	T2
6.	Linolenic acid	2.88	6.79	0.71	0.46	0.41	3.57	4.29
7.	Arachidic acid	0.54	0.64	2.14	1.95	2.14	1.78	3.31
8.	Behenic acid	2.34	0.81	5.1	4.79	5.41	3.02	1.92
9.	EPA	5.67	3.6	8.92	8.71	9.35	2.57	0.99
10.	DHA	7.42	2.17	8.02	8.98	9.29	0.82	2.42
11.	Palmitoleic acid	3.85	3.38	1.99	1.86	1.61	6.92	4.44
12.	Others	1.37	0.76	1.17	0.92	0.69	2.3	1.53
13.	total	99.93	99.93	99.94	99.94	99.95	99.95	99.94
14.	n-3	15.97	12.56	17.65	18.15	19.05	6.96	7.7
15.	n-6	26.31	36.53	19.5	20.15	19.52	10.82	18.78
16.	n-3/n-6	0.606994	0.343827	0.905128	0.900744	0.975922	0.643253	0.410011

4.5.5 Amino acid profile

Amino acid composition of Floc

Serine, Glycine, Asparagine, Glutamic acid, Alanine, and Cysteine, were found to be significantly different in T2 from T1. Histidine, Tyrosine, Arginine, Lysine, Isoleucine, Valine, Phenylalanine, Leucine found significantly higher in *insitu* compared to *exsitu*. There was no significant difference between the proline and threonine between the treatments. The NEAA, EAA and NEAA/EAA ratio was lower in *exsitu* compared to *insitu* floc.

Amino acid composition of Feed

Serine, Glutamic acid, cysteine, Proline, Histidine, Arginine, Lysine, Thysine was significantly higher in commercial feed compared to AMF feed. There was no significant difference between the Glycine, Asparagine, Alanine, Threonine, Isoleucine, valine, Leucine and Phenylalanine between both feed. NEAA, EAA, and NEAA/EAA was higher in commercial feed compared to AMF feed.

Amino acid composition of Shrimp

Lysine and Isoleucine significantly lower in control compared to other treatments. Valine had significantly higher in *insitu* followed control and *exsitu*. Phenylalanine was significantly higher in from control than treatments. The NEAA, EAA, and NEAA/EAA was found to be higher in exsitu compared to control

Table 4.11 Mean amino acid profile (g 100g-1 dried material) of two floc samples, commercial feed, biofloc feed and whole body samples of juvenile shrimp (n=3)

Amino Acid	Feed		Shrimp			Floc	
	Commercial	**Biofloc**	**Control**	***Insitu***	***Exsitu***	***Insitu***	***Exsitu***
Non-Essential Amino Acid							
Ser	6.29 ± 0.31^{d}	0^{a}	4.21 ± 0.58^{bc}	3.85 ± 0.31^{bc}	4.43 ± 0.41^{c}	4.66 ± 0.72^{c}	3.42 ± 0.42^{b}
Gly	6.92 ± 0.21^{bc}	6.60 ± 0.38^{bc}	7.13 ± 0.17^{c}	6.43 ± 0.72^{bc}	7.20 ± 0.33^{c}	6.16 ± 0.7^{b}	4.37 ± 0.37^{a}
Asp	13.24 ± 0.67^{d}	13.61 ± 0.89^{d}	8.69 ± 0.27^{b}	10.06 ± 0.21^{c}	9.62 ± 0.98^{bc}	10.07 ± 0.68^{c}	6.92 ± 0.09^{a}
Glu	19.28 ± 0.36^{d}	20.61 ± 0.66^{e}	15.63 ± 0.14^{c}	14.96 ± 0.41^{c}	15.13 ± 0.22^{c}	10.55 ± 0.08^{b}	7.43 ± 0.62^{a}
Ala	5.88 ± 0.56^{b}	6.05 ± 0.55^{b}	6.19 ± 0.48^{b}	6.38 ± 0.52^{b}	6.50 ± 0.35^{b}	6.26 ± 0.81^{b}	4.19 ± 0.05^{a}
Cys	3.36 ± 0.28^{c}	4.19 ± 0.03^{d}	2.46 ± 0.34^{b}	2.27 ± 0.38^{ab}	2.30 ± 0.58^{ab}	2.34 ± 0.19^{ab}	1.76 ± 0.11^{a}
Pro	6.97 ± 0.75^{d}	5.4 ± 0.38^{c}	3.06 ± 0.85^{b}	2.79 ± 0.15^{b}	3.00 ± 0.19^{b}	1.57 ± 0.07^{a}	1.11 ± 0.72^{a}
Essential Amino Acid							
His	3.99 ± 0.42^{d}	3.07 ± 0.57^{c}	3.02 ± 0.57^{c}	3.01 ± 0.69^{c}	$2.89 \pm 0.55b^{c}$	2.05 ± 0.21^{ab}	1.62 ± 0.9^{a}
Arg	7.26 ± 0.12^{ab}	9.98 ± 0.09^{b}	6.24 ± 0.68^{ab}	6.67 ± 0.75^{ab}	6.26 ± 0.19^{ab}	5.33 ± 0.08^{ab}	3.59 ± 0.82^{a}
Thr	4.74 ± 0.97^{ab}	5.25 ± 0.23^{b}	4.35 ± 0.9^{ab}	3.87 ± 0.1^{a}	3.74 ± 0.76^{a}	5.41 ± 0.15^{b}	3.72 ± 0.75^{a}
Lys	6.74 ± 0.75^{e}	4.43 ± 0.15^{d}	3.22 ± 0.17^{bc}	4.45 ± 0.9^{d}	4.25 ± 0.95^{cd}	2.53 ± 0.33^{ab}	1.79 ± 0.94^{a}
Ile	5.97 ± 0.96^{d}	5.54 ± 0.36^{d}	2.95 ± 0.27^{a}	3.90 ± 0.52^{bc}	4.25 ± 0.21^{c}	4.46 ± 0.13^{c}	3.22 ± 0.42^{ab}
Tyr	4.89 ± 0.5^{b}	2.98 ± 0.88^{a}	3.29 ± 0.5^{a}	3.03 ± 0.45^{a}	3.58 ± 0.97^{ab}	2.64 ± 0.18^{a}	2.92 ± 0.76^{a}
Val	5.53 ± 0.76^{c}	5.88 ± 0.55^{c}	3.82 ± 0.24^{b}	2.75 ± 0.29^{a}	4.00 ± 0.65^{b}	6.36 ± 0.14^{c}	2.89 ± 0.42^{a}
Leu	8.30 ± 0.49^{c}	7.30 ± 0.83^{c}	5.27 ± 0.35^{b}	4.87 ± 0.57^{b}	5.58 ± 0.88^{b}	5.61 ± 0.47^{b}	3.83 ± 0.23^{a}
Phe	7.80 ± 0.6^{c}	6.59 ± 0.93^{bc}	$6.74 \pm 0.95b^{c}$	5.85 ± 0.03^{b}	6.05 ± 0.95^{b}	5.90 ± 0.73^{b}	4.34 ± 0.38^{a}
Total	117.24 ± 3.89	107.58 ± 4.77	86.34 ± 3.28	85.21 ± 3.26	88.87 ± 3.17	81.95 ± 2.61	57.19 ± 1.73
NEAA	61.9714	56.5231	47.3983	46.767	48.2106	41.6425	29.2201
EAA	55.2757	51.0659	38.9463	38.443	40.6611	40.3142	27.9762
NE/EA	89.1955	90.34519	82.16814	82.20112	84.34058	96.81023	95.743

Values are mean (± SD) of the analysis

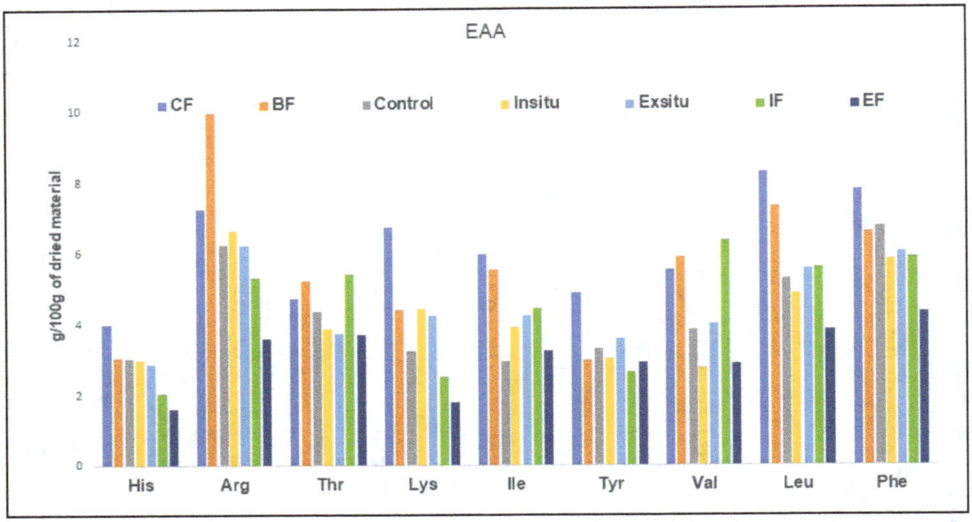

Fig. 4.9: Comparison of individual EAA (g/100g of dried material)

Table 4.10 EAA score (aa/AA)

EAA score	Feed		Floc	
	Commercial	**Biofloc**	***INSITU***	***Exsitu***
Isoleu	1	1	1	1
His	0.992584	0.832944	0.729548	0.829837
Arg	0.840625	1	0.882884	0.853912
Thr	0.880804	1	1	1
Lys	1	0.898372	0.674179	0.686172
Tyr	1	0.729352	0.846299	1
Val	1	1	1	1
Leu	1	1	1	1
Phe	0.929214	0.85609	1	1
EAAI	0.958509	0.918927	0.894737	0.923006

Apparent digestibility coefficient (ADC)

The Apparent digestibility coefficient of protein, lipid and fiber tissue of commercial feed used in T1 and C treatment was found to be around 94–99%. However, it was found little lower (98%) in the biofloc feed used in the T1 treatment.

Table 4.12 Apparent Digestibly Coefficient for the experimental diets

S. No	ADC	Protein (%)	Lipid (%)	Fiber (%)
1.	**Control**	95.24 ± 0.565	94.12 ± 0.984	99.91 ± 0.82
2.	T1	95.54 ± 0.515	94.47 ± 0.923	99.94 ± 0.88
3.	**T2**	95.68 ± 0.567	94.37 ± 0.818	99.70 ±0. 69

4.6 Floc parameters

Seventeen physical parameters of the floc were determined in this study as shown in table 4.9. There was a significant difference in mean value of floc volume between the control and treatments. The mean floc volume was found to be minimum in control (3ml/l) and higher in T1 treatment. Total solids, Total dissolved solids and Total suspended solids showed significant difference among the treatments. Higher level of TS was recorded in T1 treatment; higher TDS was observed in the T2 treatment; lower level of TSS and TVDS were found in control. The biofloc treatment exhibited gradual rise in TSS and TVDS ranged from 180 – 600 mg/l as shown in Fig 4.9 A, B, C, D.

The highest Specific surface area of floc was observed in T2 treatment (232.91 X 10^{13}) and lower in T1 (178.25 X 10^{13}). Dehydrogenase activity of floc in T2 treatment (0.0211 mg TF/ ml) was significantly higher than other treatments. Pore radius of the floc was significantly higher in control and lower in T1 treatment. Pore volume, Floc volume and Porosity of the floc was significantly higher in the T1 compared to T2. The floc concentration was relatively found be higher in T1 compared to T2

Floc volume index in T1 treatment (9.09%) was significantly lower than other treatments. In control, Floc density was significantly higher compared to other treatments. The Particle size distribution in relation to the volume of biofloc after sieving shows the dominance of biofloc size less than 20 μm during the first sampling. The floc size gradually increased in both treatments and at the end of the trial the dominance of floc size more than 20 μm was found in T1 compared to T2.

The settling velocity of the floc was higher in T2 (0.58 cm/sec) compared to T1 (0.29cm/sec) as shown in fig 4.10. settling velocity of floc gradually increased from DOC 6 to DOC 30.

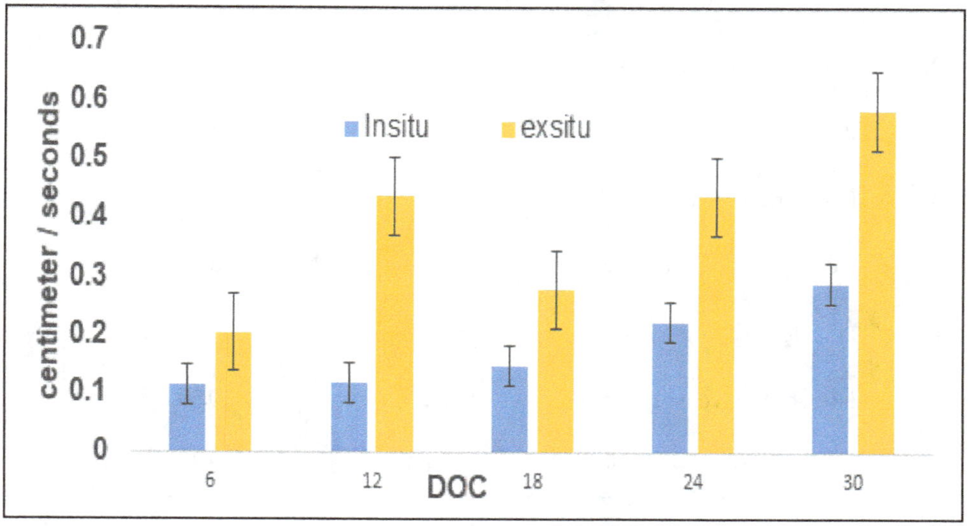

Fig. 4.10: Settling velocity

Table 4.13 : Floc parameters of treatments for the 30 days of experiment duration.

S. No	Parameter	T1	T2
1.	Floc volume (ml/l)	$11^{b} \pm 0.5$	$5^{a} \pm 0.5$
2.	TS (g/l)	21.93 ± 2.97^{a}	21.96 ± 3.69^{a}
3.	TDS (g/l)	22.09 ± 3.23^{a}	23.85 ± 4.14^{a}
4.	TSS (mg/l)	346 ± 26^{a}	386 ± 157^{a}
5.	TVS (g/l)	6.36 ± 3.09^{a}	8.45 ± 3.32^{a}
6.	TVDS (g/l)	6.73 ± 2.86^{a}	7.61 ± 2.49^{a}
7.	Size (μm)	$73.97 \text{ X } 48.56^{a}$	$36.53 \text{ X} 32.85^{a}$
8.	Shape	Irregular	Irregular
9.	Specific surface area (m^2/g)	$178.25^{b} \text{X} 10^{13}$	$232.91^{a} \text{ X} 10^{13}$
10.	Dehydrogenase activity (TF μmol/ml)	0.0197^{a}	0.0211^{b}
11.	Pore radius (μ)	$5.48^{a} \text{ X } 10^{-5}$	$17.12^{b} \text{ X } 10^{-5}$
12.	Pore volume (ml/l)	8.12	3.4
13.	Porosity (%)	12.31	29.41
14.	Floc concentration (g/m^3)	2.88^{a}	1.6^{b}
15.	Floc density index (ml/g)	38.19^{a}	31.25^{a}
16.	Floc density (mg/cm^3)	0.26^{a}	0.32^{a}
17.	Floc settling velocity (cm/s)	0.29	0.58
18.	Floc volume index (ml/g)	9.09^{a}	20^{b}

Values in the same row with different superscripts differ significantly ($P<0.05$) for each parameter. One way ANOVA was used following Duncan multiple range test in SPSS

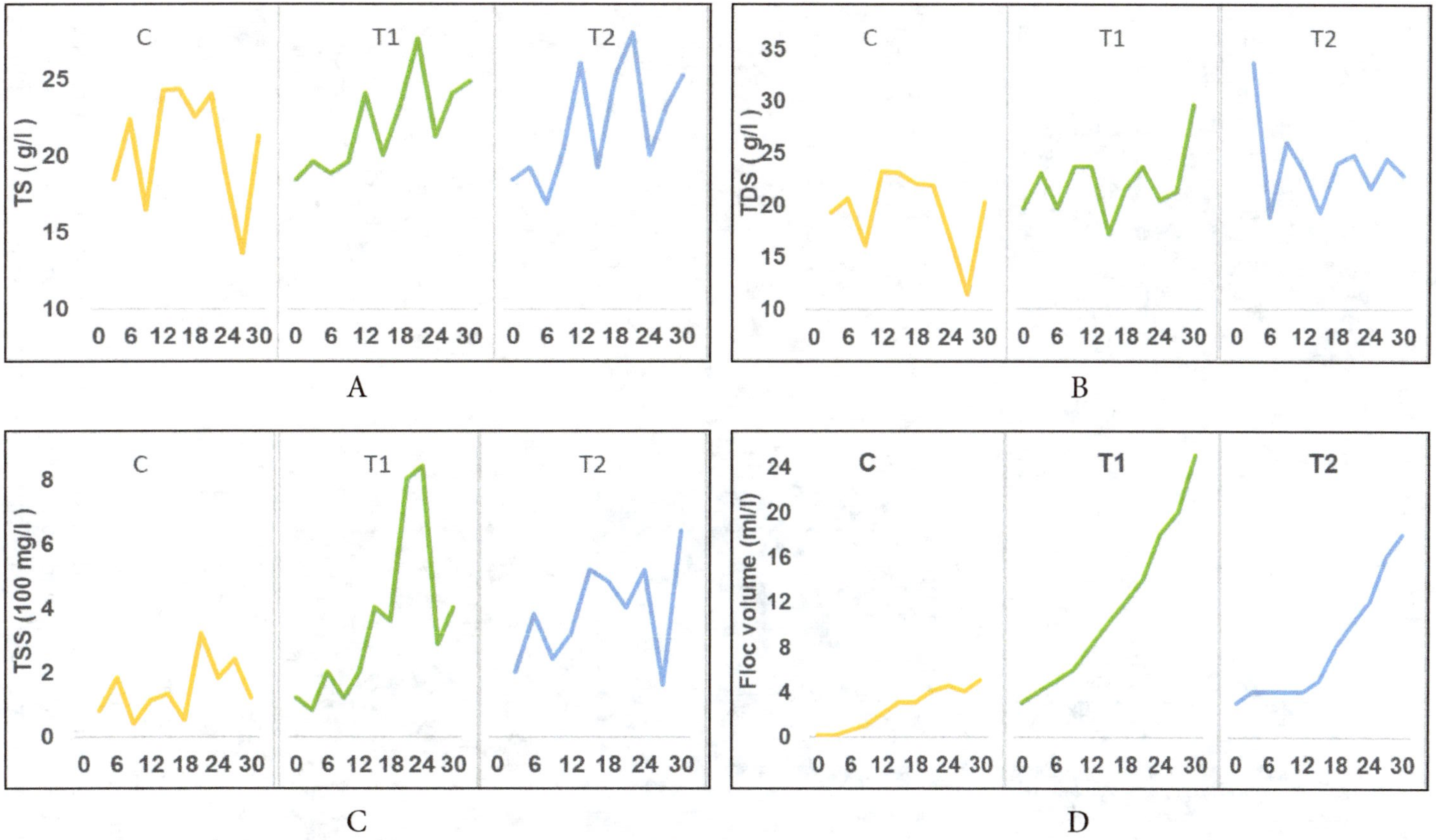

Fig. 4.11: Floc parameters of treatments for the 30 days of experimental duration.

Plate – 5 Floc images

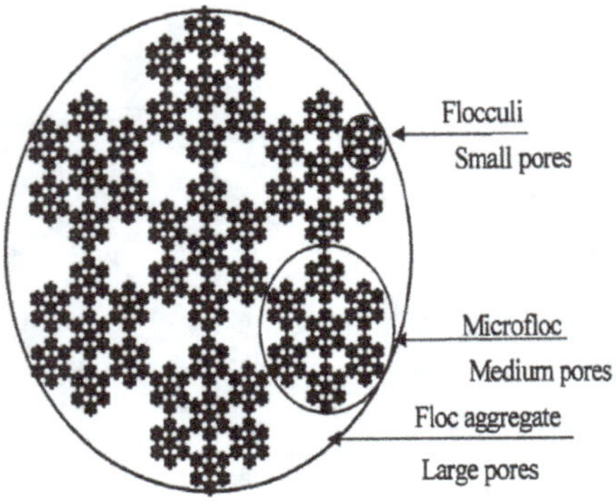

5A - Theoretical structure of floc

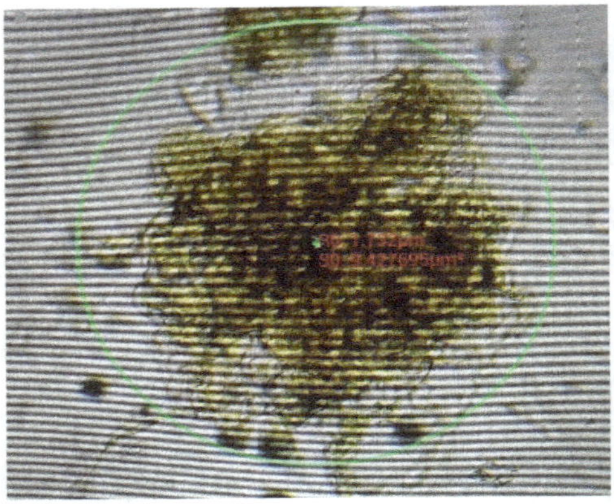

5B – Floc image under digital microscope

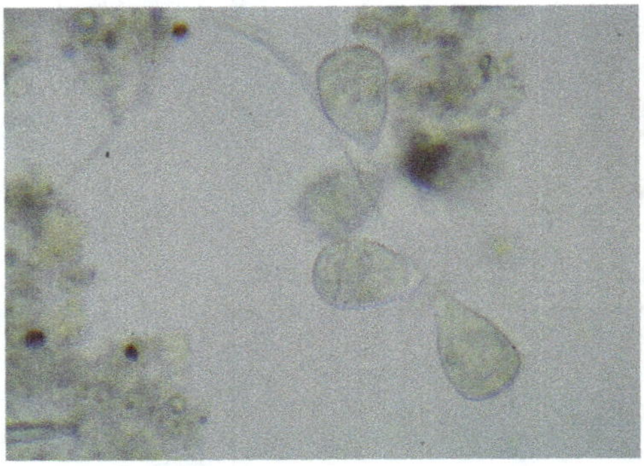

5 C – Young AMF

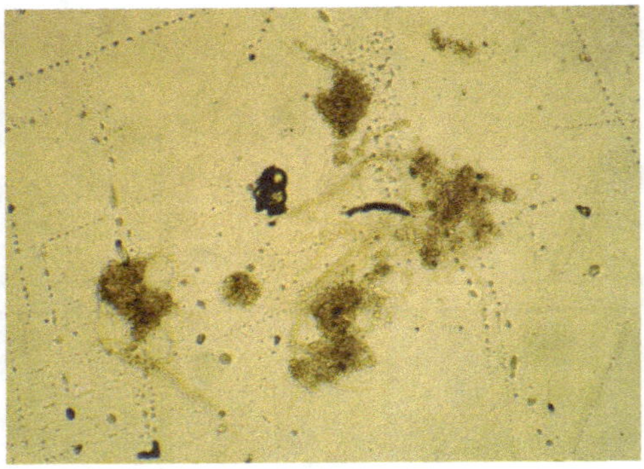

5D – AMF formation

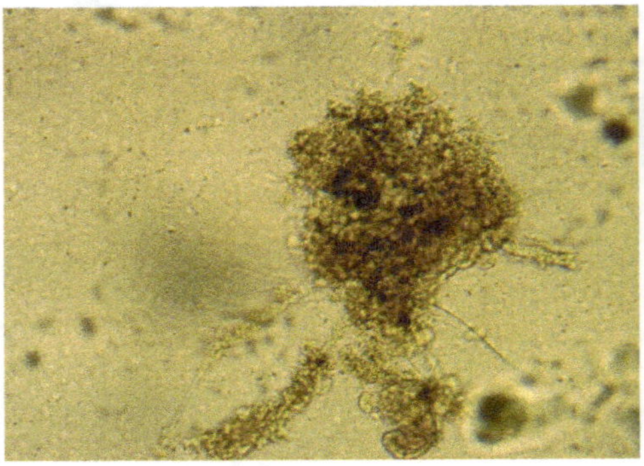

5 E – Maturing stage AMF

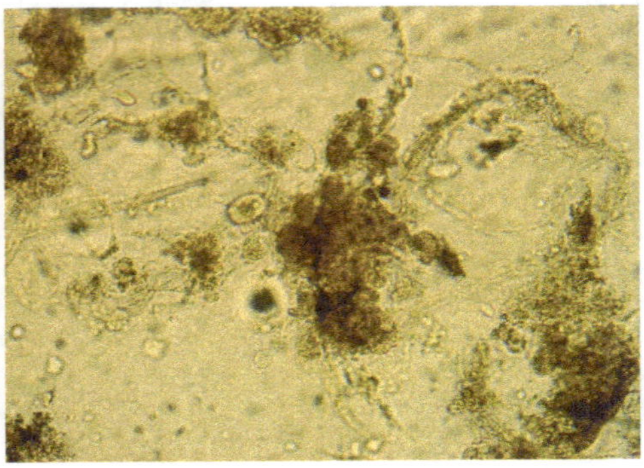

5F – Matured stage AMF

4.7 Economic viability

The benefit cost ratio of the AMF driven nursery rearing of shrimp in raceway was above 1.00 in indicating that adoption of AMF based nursery culture is an economically viable venture. In comparing with total outcome, shrimp culturing in AMF based *insitu* and *exsitu* systems showed higher yield and similar benefit cost ratio.

Table 4.13 Economic viability of AMF nursery shrimp culture

S. No	Fixed cost items	Cost (in lakhs)
1	Raceways	**0.8**
2	Green house structure	**0.75**
3	Blower(5hp)	**0.90**
4	Generator	**1.0**
5	Rapid sand filter	**0.4**
6	Bore well and air line	0.3
7	Depreciation (10%)	**0.41**
8	Miscellaneous	**0.33**
Total		4.9

S. No	Variable cost	Cost (Rs)		
		C	T1	T2
1	Shrimp seed (0.35 Rs/seed)	**15,750**	**15,750**	**15,750**
2	Feed	**2,400**	**2,400**	**1,890**
3	Power consumption	**6,666**	**6,666**	**6,666**
4	Fuel	**666**	**666**	**666**
5	Labor	**6,666**	**6,666**	**6,666**
6	Interest on investment	**16,333**	**16,333**	**16,333**
7	miscellaneous	**5000**	**5000**	**5000**
Total		53,481	53,481	52,971

S. No	Parameters	Yield		
		C	T1	T2
1	Survival (%)	80	89	89
2	Possible crops per year	6	6	6
3	DOC	30	30	30
4	Total no of seeds produced	36,000	40,000	40,248
5	Cost of seed selling price (Rs / shrimp)	1.75	2	2
6	Total returns (Rs)	63,000	80,000	**80,500**
7	Net returns (Rs)	9,519	26,519	27,525
8	Benefit cost ratio	1.17	1.5	1.5

CHAPTER - 5

Discussion

The primary aim of the experiment was to study the impact of *insitu* and *exsitu* AMF driven culture in nursery rearing of *P.vannamei*. The present study confirmed to large extent on the beneficial effects of AMF in improving water quality through removal of toxic nitrogen species by increasing feed utilization, growth performance of shrimp and stimulating digestive enzyme activities. The floc composition also enhances the biosecurity and health management of shrimps with possible probiotic effect in zero water exchange system.

5.1 Water quality parameters

Water quality parameters recorded during the experiment were in the optimal range of nursery rearing of *P.vannamei* indicating that the experimental condition was conductive for the growth of the animal (Felix, 2015b). Dissolved oxygen observed in this study (5.07 mg L-1) was sufficient for good survival and growth of *P.vannamei* (McGraw et al., 2001).The sufficient DO level in all the treatments during the experiment period was attributed due to continuous aeration. Water pH in all the treatments were found within the ideal range of 7.0-8.8 (Van wyk et al., 1999). The water hardness ($CaCO_3$) and alkalinity levels were above the recommended levels of 1500 and 110 mg/l respectively (Boyd et al., 2002) and might therefore, have provided the physiological conditions allowing shedding and proper formation of the exoskeleton, promoting growth and survival of organisms (McGraw and Scarpa, 2003).

The assimilation of nitrogen was also evidenced from the significantly higher crude protein content observed in the AMF (T1 and T2) treatments. In the present study use of distillery spent wash to maintain the C/N ratio 10:1 was efficient in maintaining low level of nitrogen compounds (NH_4-N, NO_2-N, NO_3-N) and orthophosphate (PO_4-P) without substantial increase in COD and BOD (Avnimelech, 1999). This carbon source was efficient in ammonia assimilation and thus produced good quality floc (Felix et al., 2015; Menaga et al., 2017; Liu et al., 2017). The presence of high concentration of NO_3-N in AMF treatments indicates the occurrence of nitrification processes in the culture systems. While NO_2-N concentration in AMF treatments seems to be relatively stable, the opposite was observed in the control which could be explained by the high rate of nitrification process in these treatments. The nitrogen dynamics and reduction in TAN concentration confirms the conversion of organic nitrogen by using distillery spent wash as a substrate and this finding was consistent with those of previous studies (Liu et al., 2017; Felix, 2015; Raj A et al., 2017; Lu et al., 2012; Luo et al., 2013; Yao et al., 2013;). Thus the carbon induced nitrogen management in aquaculture system will go a long way enabling enhanced fish production with reduced input cost (Felix et al., 2015).

5.1.1 Microbiology parameters

The increment of total vibrio counts in AMF treatments (T1 and T2) was coherent with the early findings. Even though biofloc was thought to have a probiotic effect on shrimp and stimulate their immune system (Schryver et al., 2008; Crab et al., 2012; Ekasari et al., 2014; Kim et al.,2014) Vibrio outbreaks such as in this trial are still common in these systems. These outbreaks usually occur in conjunction with a stressor (Sung et al.,2001, Ekasari et al., 2014). Although mean water quality variables were within acceptable levels for shrimp in this study, ranges included high that could potentially compromising shrimp health and increasing susceptibility to infections (Cheng et al., 2003). The coefficients of variation of total heterotrophic bacterial concentrations indicated a slightly higher variability in the AMF treatments than in the control cultures. Concentrations were in the same order of magnitude than the mean value found by Kim et al. (2014) in BFT-managed shrimp cultures, but one order of magnitude lower than those determined in BFT-managed shrimp cultures by Burford et al. (2004).

5.2 Growth parameter

In the present study, *P.vannamei* reared under AMF using distillery spent wash showed a better performance in terms of mean body weight, SGR, FCR, PER, survival and yield compared to control. The results were in agreement with the findings of Xu and Pan, (2012);Serra et al. (2015). However, significant difference in PER was found between T1 and T2 and the efficiency was higher in T2 due to increased weight gain of the culture animals by the synergistic effects of the improved water quality and higher heterotrophic bacterial community.

Inclusion of AMF at 25% level as a dietary ingredient in shrimp diet (T2) found to improve the growth performance and relatively higher survival of *P.vannamei* (Ju et al 2008b; Kuhn et al 2009, 2010). This difference was mainly due to the composition of biofloc, as they are the rich source of many bioactive compounds such as carotenoid, chlorophylls, phytosterols, bromophenols, amino sugars (Ju et al., 2008a) and antibacterial compounds (Crab et al., 2010). The presence of microbial compounds, unknown growth factors, probiotic micro-organisms in the AMF might have resulted in significantly higher growth rate and better FCR in shrimp fed with AMF incorporated diet. FCR in T1 was also less compared to control and agrees with the findings of Avnimelech et al. (1994) and Panjaitan, 2004 who found that lowering feed application up to 30 % of conventional feeding ration, did not lower shrimp growth, probably due to the partial replacement of feed by the microbial flocs. AMF treatments resulted in higher survival due to the provision of essential nutrients, such as amino acids, vitamins, and minerals present in the flocs (Decamp et al., 2002).

5.2.1 Quantification of RNA - DNA ratio in muscle

RNA content in T2 and control was high in the initial 15 days and at the end of the sampling it was found in the decreased level compared to T1. As somatic growth of animal is directly related to the protein synthesis in tissues. RNA-DNA ratio is directly correlated with protein synthesis as RNA is the precursor for the protein synthesis. RNA-DNA ratio was significantly higher in control than treatment due to unfavorable water quality parameters affecting the growth and survival. T1 exhibited higher RNA-DNA ratio and this may be due to stress caused by the increased floc volume in T2 (Wilder and Stanley 1983). The changes in nucleic acids implicates changes in growth performance of the animal and this parameter as indicator for growth and nutritional status of animal was clearly observed between T1 and T2 (Tewery and Patra, 2008). The result obtained from present study was within range of 0.20 to 0.32 and agrees with of findings of Moss, (1994).

5.3 Biochemical composition

5.3.1 Biochemical composition of AMF

The proximal composition or nutritional value of biofloc changes according to the culture conditions and is closely related to the microflora of the biofloc environment (Ju et al., 2008a; Xu et al., 2012). Thus Crab et al. (2010a), unlike in our study, found a lower lipid content (2.3 to 5.4 % of DM) and only a trace of n-3 fatty acids (0.4 to 0.7 mg g^{-1}).

The lower crude protein level (23%) noticed in harvested biofloc compared to earlier findings (Crab et al., 2010; Ekasari et al., 2010) may be due to difference in the bacteria taking part in floc formation (Rittmann and Mccarty, 2001). The decreased level of protein might be contributed by the microbial floc composition as the microbiota of the floc is likely to affect the protein content of the biofloc (Ju et al., 2008a).

Ash and crude lipid contents in the present study averaged 34.75 and 0.95% respectively, were also close to the values reported by Wasielesky et al. (2006). The gross energy on dry matter basis 2550 kcalg^{-1} can be considered as appropriate in fish nutrition especially for herbivorous/omnivorous fish and shrimp species (Webster and Lim, 2002).

In the microbial food web, micro algae and zooplankton are the rich source of lipids compared to bacteria (Zhukova and Kharlamenko,1999). Out of 11 fatty acids detected in AMF, palmitic acid, linoleic acid and oleic acid were the dominant fatty acids, and omega 3 fatty acids detected in trace level. This is in consonance with the findings of Crab et al., (2010) and the lower level of PUFA in biofloc may be due to the dominance of heterotrophic bacteria compared to PUFA rich microalgal community (Meyers and Latscha,1997).

The essential fatty acids for shrimps found in biofloc aggregates were dominated by EPA (0.99–2.57 % of the dry matter), DHA (0.82–2.42% of DM) and ARA (1.78–3.31% of DM).The dietary requirements for these 3 essential fatty acids for *Penaeus japonicas, Penaeus monodon* and *Penaeus vannamei* were found to be 0.5 to 1.1 % (EPA and DHA) and 0.5% (ARA) respectively (NRC, 2011).The similarity of results with the early findings (Menaga et al., 2015) using distillery spentwash marked its significant application in producing quality floc for fish meal replacement in the shrimp diet.

5.3.2 Biochemical composition of feed

The proximate values was not different from each other between any of the diets used in T1 andT2, which were concededly optimal for shrimp culture (Davis et al., 1993; Cuzon et al., 2004; Fox et al., 2006). This allows direct comparison between the diets with and without microbial floc. Since the other nutrients are either constant and not limited in the control and treatment diets, the difference in growth rate between the treatments was due to the replacement of fish meal by AMF meal ingredient in T2. The average ash content found in the range of 7-12 % was lesser than the earlier findings, where the ash content ranged from 7 to 32 % on the DW (Tacon et al.,2002; Azim et al., 2007; Ju et al.,2008a). Furthermore, Tacon et al., (2002) suggested that the high ash content in the bio-flocs probably related to the presence of acid insoluble oxides and mixed silicates.

The percentage of relatively higher amount of Lipid in biofloc based diet in T2 showed the similar trend of fatty acid profile as that of commercial feed diet. Presence of Oleic acid, Linoleic acid and Linolenic acid were higher than the commercial feed, however decreased level of EPA and DHA were observed in the T2 diet. Nevertheless, it was within the recommended level of nutritional requirement of shrimps (NRC,2011). Interestingly this didn't affect the shrimp growth probably due to the presence of biofloc in the culture environment as the fatty acids requirement of aquaculture animals could be met by other feed resources (Rinchard et al., 2007; Crab et al., 2010b).

5.3.3 Biochemical composition of shrimp

The biochemical composition of the nutrients in shrimp body shows similar trend in all the treatments except the crude protein level. Although few differences among the protein content in the shrimp body tissue do not cause large difference in survival between T2 and T1, a possible balance between dietary nutrients and biofloc nutrients might have influenced this relative change (Toledo et al., 2016). There was no striking difference in the fatty acid profile of shrimp whole body tissues between the treatments. The greater presence of C16:0 and C18:1 n-9 in the samples of hepatopancreas of *P vannamei* was also observed by Emerenciano et al. (2012). As expected the high amount of palmitic acid (C16: 0) and oleic acid (C18:1 n-9) was observed in hepatopancreas of shrimp. was expected because, in general, fatty acids can be synthesized *de*-novo by all organisms, including fish and crustaceans, by the addition or removal of 2-C acetyl units (D'Abramo et al., 1997; Glencross, 2009). The shrimp possess the enzyme system delta-9 desaturase that can convert these saturated fatty acids in monounsaturated forms which contain a double bond (D'Abramo, 1997; Sargent et al., 2002). We observed low levels of PUFA in shrimp whole body that can reinforce the idea that these long-chain fatty acids are not synthesized or may have been used for other functions such as membrane structure and immune system (Glencross 2009).

5.3.4 Amino acid profile

Total EAA in feed and biofloc were very similar to those previously reported in commercial (Morenoarias et al., 2018; Ju et al., 2008) and experimental feeds with biofloc meal partially replacing fishmeal (Kuhn et al., 2010, Bauer et al., 2012

Morenoarias et al., 2018), as well as fed with vegetable meals in order to reduce fish protein (Jatobá et al., 2014).

Considering the A/E ratios, in the present study methionine and tryptophan were not detected in both feeds and floc sources, this was also observed previously by Ekasari et al. (2014). It has been stated that methionine limiting in the diet can cause reduced growth and low feed efficiency (National Research Council, 2011), but these effects were not observed in our study. Probably, because of the presence of biofloc in the AMF based treatments might have complemented the shrimp requirements.

The EAAI profile indicates that arginine and lysine had a low concentration in both feed sources (AMF insitu and exsitu). Comparing to Ekasari et al. (2014), our EAAI values of feed supplemented with AMF contains adequate levels for shrimp nutrition. The main reason for these results were based on the microorganism's ability to biosynthesize the 20 primary amino acids (National Research Council, 2011). It is well known that the nutritional value of bioflocs can be influenced by the developmental stage of microbial biofloc, whether autotrophic, heterotrophic, or chemoautotrophic, as well as the adopted management plan.Also, the heterotrophic phase provides more nutritional value (Avnimelech, 2006; Ebeling et al., 2006; Wasielesky et al., 2006; Taw, 2010) than the others. The levels of crude protein in the biofloc were higher than those observed by other authors (Chamberlain et al., 2001; Tacon et al., 2002; Williams et al., 2005; Wasielesky et al., 2006; Taw, 2010; Jatobá et al., 2014), who recorded values of crude protein below 20.0%.This highy influenced the enriched aminoacid profile of the shrimps cultured in T1 and T2.

5.4 Digestive enzyme activity

Biochemical composition of the diet plays an important role in the digestive enzyme profile in the shrimp (Gamboa-delgado et al., 2003). In the present study dietary supplementation of AMF in T2 significantly improved ($P<0.05$) the specific activity of the digestive enzymes like protease, amylase, lipase and cellulase in the treatment groups compared with control. The increased digestive enzyme activity in *exsitu* treatment (T2) might have enhanced the digestion and nutrient absorption and recorded significantly higher ($P< 0.05$) growth rate in this treatment. A previous study by Xu and Pan, 2012 in biofloc based system reported the similar results in *P. vannamei.* This may also be due to the interference of live microorganisms attaching to the biofloc after transition through the stomachs in resident intestinal microflora balance, which plays an important role in the production or secretion of digestive enzymes (Harris, 1993; Moss et al., 2000; Xu and Pan, 2012). Thus, the biofloc could influence digestive processes of the shrimp by increasing exogenous enzyme, inducing endogenous enzymes and altering intestinal microflora balance. In the present study, the bacterial community of the AMF might have improved the assimilation of AMF meal that was included in the T2 diet. The present study also disagrees with the findings of Ajiboye et al. 2012; Kiessling and Askbrandot, 1993 in compliance with the addition of microbial products at higher level (> 4%) reduce the feed pellet stability and digestibility.

Similarly, enhanced level of digestive enzyme activity has been reported in fish and shrimp fed with probiotic, microalgae and periphyton supplemented diet (Anand et al., 2013b). Lara-flores et al, 2003; Ziaei-nejad et al. 2006 reported the presence of microbial compounds in the biofloc supplemented diet might have stimulated the production of endogenous enzyme by the shrimp visceral mass. Despite of the higher ash and fiber level recorded in the T2 diet, the cultured animal showed higher specific enzyme activity compared to other treatments. Nevertheless, this may be due to the fact that the shrimp possess a suite of a digestive enzyme capable of hydrolyzing a variety of substance and several factors have been implicating in altering digestive enzyme activity including diet (Lee et al., 1984).

5.5 Extracellular enzyme activity

The collected flocs in AMF treatments showed significant difference between the treatments. The higher activities of protease, amylase and lipase indicating that relevant extracellular enzymes could be secreted by the microorganisms attached to the AMF. Secretion of cellulase enzyme was completely absent in control, perhaps trace level activity was recorded in AMF treatments. This clearly shows the presence of diversity and activities of microbial consortia present in the AMF treatments. This research study also finds the differences in the expression of enzyme specific activity between T1 and T2. As the presence of heterotrophic bacteria was relatively higher in T2, this attributes to the secretion of extracellular enzymes at significant level compared to T1. With this findings, we cannot distinguish between enzymes activities synthesized by the shrimp and that produced by the microorganisms. As it can be speculated that extracellular enzymes attaching to the biofloc can release to the digestive tracts after being ingested by the shrimp. Those acquired enzymes would present, at least, a portion of the contribution to the specific enzymes activities as a significant increase in protease, amylase and lipase activities in the hepatopancreas of the T1andT2. Perhaps, the presence of the AMF may also in some way stimulates endogenous enzymes produced by the shrimp (Moss et al., 2001). This can be further demonstrated by the fact that the activities of digestive enzymes such as proteases, amylases, lipase and cellulase of hepatopancreas, stomachs and intestine, not including exogenous enzymes produced by the AMF, tend to increase in the AMF treatments.

5.6 Apparent digestibility and Bioavailability

In the present study, both types of feed used in T1 and T2 displayed a very good digestibility for crude protein, crude lipid and crude fiber. Till date, only one study on bioavailability of crude protein and crude lipid was carried out at pilot scale by using biofloc meal incorporated diet by Neto et al. (2015) and ours would be the first study to be conducted in raceways. The outcome of our research study clearly shows the suitability of using biofloc incorporated ingredient in shrimp diet as the ADC for crude protein (95%), crude lipid (94%) and crude fiber (99 %) was within the recommended level of 58-97% according to NRC,2011. This result suggests that biofloc may have a growth promoting effect on *P.vannamei*. It is possible that this effect is not directly associated with the protein present in the biofloc, but another nutrient, possibly macro-(calcium, phosphorus, potassium and magnesium) and (or) micro minerals (copper, iron, manganese, and zinc) as suggested by Neto et al. (2015).

5.7 Floc parameter

Although the physical properties of flocs are probably the most important factors affecting the nutritional composition (Ekasari, 2016) as well as the sludge formation, a detailed study on the physical characteristics of floc in biofloc systems has not been studied much. This study would be the first of its kind to determine the physical characteristics of floc in both *insitu* and *exsitu* treatments.

The TSS observed in this study T1 and T2 was within the recommended level of <500 mg/l for penaeid shrimps (Samocha et al., 2007). Several authors have indicated a similar trend of concentration of TSS which is beneficial to the shrimp and to the system stability (Schryver,et al., 2008). The levels of TSS in T2 treatment were significantly higher than the control. These differences stemmed from the higher levels of non-digestible components in AMF treatments. Relation between non-digestible components of feed and solids increase in biofloc system was previously suggested by Lopez-Elias et al. (2015).

The floc density recorded at the end of the trial was higher in T2 compared to T1. The low density of floc aggregates may be attributed to the density of filamentous bacteria in the floc (Barbusiński and Kościelniak,1995).

The floc size in AMF treatment at the initial sampling agrees with the findings of Ganczarczyk, (1996) who estimated the size of microbial clusters in both microbial flocs and biofilms to vary from 6 – 12 μm. Perhaps floc size (>100 μm) at the end of the experiment in T1 was similar to the findings of Ekasari et al. (2014).

Floc porosity was higher in T1 compared to T2 and this finding agrees with the empirical correlation defined by stokes law. This may be due to the existence of increased floc size in T1 than T2. The increase in floc size in T1 might have been affected due to floc volume and mixing intensity. This difference is mainly because of the size of the floc which is determined by the energy input and mixing intensity.

Dehydrogenase enzyme activity have been widely used to assess the metabolic state of microbial community and responsible for electron transport in the water environment. Although research into this theme doesn't exist so far in aquaculture systems, the dehydrogenase activity is correlated with floc volume, as the amount of floc volume is lesser in T2 and this may lead to the higher dehydrogenase activity in T2. Strong correlations are undoubtedly connected with the fact that the metabolism and the survival of microorganisms are also strongly impacted by the availability of water (Uhlirova et al., 2005) which is essential for microbial activity. Consequently, low water availability in T1 can inhibit microbial activity by lowering intracellular water potential, and thus reducing hydration and dehydrogenase enzymes activity (Wall and Heiskanen, 2003).

Floc density was lower in T1 than T2, this may be probably due to higher floc porosity. It is evident that smaller flocs have higher densities than larger flocs, which is in line with the findings by Lee et al., (1996). This research finding also correlates with the findings of floc porosity according to Huang, (2004) as floc size increases its density decreases.

The specific surface area of flocs in T2 was higher than T1, this is mainly due to the presence of smaller floc size in T2. This was coherent with the findings of Harris, (1993). The results indicated that the floc size reduction effect was a dominant one in determining specific surface area. The specific area of the floc is related to pore radius and larger the surface area indicating that the flocs are highly porous in nature. This porous nature also helps in adsorbing particulate matter and high diffusion rate of nutrients and oxygen (Coackley,1985).

Mean floc pore radius is determined with the characteristics describing the fine floc physical structure. Since little is known about them so far, the study on the pore radius and pore volume have a correlation with specific surface area of floc (Broekhoff and Liensen, 1970). With increase in specific area the increase in pore volume was observed along with decreased pore radius. As the floc size in T2 was smaller, the pore radius in T2 was relatively smaller and attributes to the porous nature of the flocs.

The pore volume in T2 was lesser which leads to increase in the floc concentration and this may be due to the presence of finer small sized flocs. The Floc volume index defined as the volume that a certain mass of flocs in the water(Avnimelech,2016). The FVI was higher in T2 than T1 which indicates the better utilization by the shrimps in improving its feed conversion ratio. Flocs with a higher FVI give the aquaculture organisms more opportunity to filter the flocs from suspension before they finally sediment to the bottom of the ponds and are lost as feed (Avnimelech, 2006). The higher FVI also indicates the age of floc. The flocs with low FVI have good settling and compaction characteristics. As FVI and floc density index (FDI) are inversely proportional to each other, lower FDI was found in T2 and has lesser compaction compared to T1.

The floc volume was maintained as per the recommended level (<15 ml) (Avnimelech, 2009) for shrimp culture in both the treatments. FV-15 is commonly used to evaluate the sedimentation characteristics and quantity of biofloc (Avnimelech, 1999,2012; Lu et al., 2012). The FV-15 only increased in the treatment groups during the initial experimental period, which was probably due to increasing quantities of biofloc due to microbial proliferation.

The settling velocity in T2 was higher, due to increased floc size when compared to T1. The settling velocity mainly influenced by floc volume and floc shear rate caused by the intensive aeration. The same trend was recored in the previous studies (magara et al., 1986).

With regard to shapes of microbial biofloc, irregularity is the major feature, even though many investigators have proposed numerous common shapes for the flocs. As a floc can be observed directly from only one direction, it is difficult to determine its three dimensional shape.

5.8 Economic analysis

Nursery rearing of *Penaeus vannamei* in raceways was found economically feasible

(Felix et al., 2015) under biofloc technology. The higher benefit cost ratio recorded in biofloc based diet clearly indicates that the rearing of *P.vannamei* will be more profitable compared to commercial feed based culture. It is possible to raise the production level to many folds as this system ensures high survival. Further, the quality of culture practice adopted in such animals fed with biofloc diet are at its high level. The hi-tech practice, using bioremediation techniques in such system have proved to be the successful formula for enhanced shrimp production. Its technically feasible, economically viable and it is also eco-friendly for adoption (Felix, 2014). The economic analysis of profitability indicated that AMF feed developed would be commercially viable in a super intensive, biofloc dominated, zero-exchange system. The total production cost per kg of harvested shrimp was more in control compared to T1 and T2, which is probably associated to higher amount of used feed and variable costs (e.g. use of probiotics, lime, fertilizers, etc.,). However, between the treatments T2 showed relatively low production cost than T1. The net returns above all costs was lower in T2 resulting in a longer payback period, and lower net present value and internal rate of return for this feed. Based on these results, we can conclude that the lower-price biofloc based feed financially viable than expensive commercial feed under aerobic microbial floc driven *P.vannamei* culture.

CHAPTER 6

Summary and Conclusion

Aquaculture is going to be an essential fish protein supplementing sector in future for the emerging global population. The current challenge for aquaculture is to develop a sustainable production technology which can utilize underutilized resources to increase productivity. Research is needed to find an innovative way to improve the present farming system in underutilized brackish water areas by adopting new technique or species which can grow well in that environment. The present study was undertaken to assess the bioavailability of crude protein and lipid from AMF driven nursery culture of *P.vannamei.* The shrimp PL was reared in a raceway system with a stocking density of 1500PL/m^2 for the period of 30 day in three treatments as clear water system fed with commercial feed (control), *insitu* based AMF system fed with commercial feed (*insitu or* T1) and *exsitu* based AMF system fed with AMF feed (*exsitu* or T2). The salient findings of the present study have been summarized as follows:

Water quality parameters during the rearing period were ideal for the growth and good health of the shrimp. Toxic nitrogenous species were on a tight rein with zero water exchange for a period of 30 days. Total heterotrophic bacterial count was higher in AMF based treatments with the influence of proper maintenance of C:N ratio at 10:1.

Extracellular enzyme activity (protease, amylase and lipase) was higher in the AMF treatments and lower in control. Protease activity (U μg protein^{-1} min^{-1}) was higher in *exsitu* treatment (49.697 ± 2.55) followed by *insitu* 52.761 ± 0.5622) and control (3.5498 ± 0.347). Lipase activity (U μg protein^{-1} min^{-1}) was higher in *exsitu* (726.392 ± 104.61) subsequently insitu (481.7535 ± 98.337) and control (263.608 ± 9.3199). Amylase activity (U μg protein^{-1} min^{-1}) was higher in *exsitu* (43.3936 ± 1.22) afterward *insitu* (38.619 ± 0.59) and control (5.811 ± 0.0616). Cellulase activity (U μg protein^{-1} min^{-1}) was higher in exsitu (0.487 ± 0.001) and lower in control (0.004 ± 0.00035).

Specific activity of digestive enzyme (protease, amylase, lipase and cellulose) recorded at shrimp digestive organ (hepatopancereas, stomach and intestine) showed a significant difference between treatment and digestive organs,. Amylase, lipase and cellulase activity was higher in T2 followed by T1 and control. Protease enzyme activity was higher in T1 treatment compared to others in stomach and intestine. The bacterial community of the AMF also improved the digestive enzyme activity by inducing the secretion of extracellular enzymes from AMF meal. The floc parameters such as floc size, floc porosity, floc volume and mean floc pore radius was higher in insitu treatment whereas dehydrogenase activity, floc density, specific surface area and floc volume index, settling velocity was higher in T2 treatment.

Growth parameters obtained from the study clearly indicates that shrimp reared in *exsitu* and *insitu* system exhibited higher growth compared to control treatment. Biofloc reared shrimps recorded the significantly higher SGR (% day^{-1}), in *exsitu* (13.15 ± 0.12) treatment followed by *insitu* (13.14 ± 0.15) compared to control (13.05±0.2). No significant difference was observed between *insitu* and *exsitu* treatments. The lowest FCR was recorded in the *exsitu* (1.044± 0.02g) treatment followed by the *insitu* (1.047 ± 0.02) and control (1.076 ± 0.03). Among treatments, the highest FER was recorded in *exsitu* (0.9571 ± 0.014) subsequently by *insitu* (0.9544 ± 0.01) and control (0.929 ± 0.01).PER of the present study indicates that *exsitu* (0.031 ± 0.01) system showed improved performance followed by *insitu* (0.023 ± 0.01) and control (0.02813 ± 0.01).

A significant difference was observed between all the treatments. At the end of the trial highest RNA-DNA ratio was recorded in control (0.2286 ± 0.0040) compared to *insitu* (0.20 ± 0.0084) and *exsitu* (0.20 ± 0.0054) treatments influencing its growth parameters.

Whole body composition of shrimp reared in different treatment showed the significant difference in protein content and the total ash. Shrimp raised in the *exsitu* has lower protein (42.78%) and gross energy (4216 $kgcal^{-1}$); a higher level of total ash (16.6%) in control. The (0.9 – 0.97) n-3/n-6 ratio was found at higher level in the commercial feed (0.6) compared to AMF feed (0.34). In amino acid profile, there was no significant difference between the shrimp muscle in all treatments, a significant difference was observed between the commercial and biofloc feed. *Insitu* biofloc had a significantly higher level of AA content compared to *exsitu* system. The total AA content ranged from 57 – 81g/100g, 117-107 g/100g, and 85-88g/100g with respect to floc, feed and shrimp. Percentage of EAA levels in floc sample possessed approximately half of the total AAs content, suggesting that these materials may be a good supplemental source of EAA for shrimp. The biofloc feed and commercial feeds had EAA levels of 52.8% and 52.5%. Aerobic microbial floc feed had a higher value of apparent digestibility (94-99%) and higher bioavailability which is suitable for the shrimp diet. In an economic point of view, both *insitu* and *exsitu* treatments were suitable for the intensive culture system as they Yield the higher benefit cost ratio.

Over all present study, states that heterotrophic bacteria in Aerobic Microbial Floc system can maintain optimum water quality for a nursery culture under zero water exchange by assimilating the nitrogenous compounds within the system and convert them to useful protein. The nutrient rich microbial biomass can also serve as food for the culture organism. *P.vannamei* could effectively harvest and assimilate the microbial floc and confer better growth performances (weight gain, growth rate per day, FCR, FER, PER, SGR, Yield), enzyme activity, and survival under AMF based *insitu* and *exsitu* feeding system.

6.1 Further studies and recommendation

Further research on the means of growing bioflocs in a way that harvesting efficiency will rise and for selecting the more favorable microbial community, will probably leads to a sophisticated biotechnological development of biofloc.

A large part of the work on the quantitative and qualitative characteristics of floc to enhance the better utilization of animals has been to be worked.

Biofloc made with specific microbial communities can be developed for formulating shrimp diet which might be nutritionally better than other.

References

A.I.D.A., Annual Report, All India DistillersAssociation and Sugar Technologists Associationof India, New Delhi, 2007.

A.O.A.C., 1995. Official methods of analysis of AOAC International. *Arlington, Va.: AOAC Intl. pv (loose-leaf).*

Abdel-Tawwab, M., Mousa, M.A. and Abbass, F.E., 2007. Growth performance and physiological response of African catfish, Clarias gariepinus (B.) fed organic selenium prior to the exposure to environmental copper toxicity. *Aquaculture,* 272, 335-345.

Ajiboye, O.O., Yakubu, A.F. and Adams, T.E., 2012. A perspective on the ingestion and nutritional effects of feed additives in farmed fish species. *World Journal of Fish and Marine Sciences,* 4, 87-101.

American Public Health Association, APHA. 2005. *Standard Methods for the Examination of Water and Wastewater. 21st ed. American Public Health Association, Washington DC, 1220p.*

Anand, P.S., Kohli, M.P.S., Roy, S.D., Sundaray, J.K., Kumar, S., Sinha, A., Pailan, G.H. and kumar Sukham, M., 2013. Effect of dietary supplementation of periphyton on growth performance and digestive enzyme activities in *Penaeus monodon. Aquaculture,* 392,59-68.

Anand, R., Gill, K.D. and Mahdi, A.A., 2014. Therapeutics of Alzheimer's disease: Past, present and future. *Neuropharmacology,* 76, 27-50.

Anand, S., Sudhayam, P., Kumar, S., Kohli, M.P.S., Sundaray, J.K., Sinha, A., Pailan, G.H. and Dam Roy, S., 2017. Dietary biofloc supplementation in black tiger shrimp, Penaeus monodon: effects on immunity, antioxidant and metabolic enzyme activities. *Aquacult. Res., 48,* 4512-4523.

Arnold, S.J., Coman, F.E., Jackson, C.J., Groves, S. A., 2009. High-intensity, zero water exchange production of juvenile tiger shrimp, Penaeus monodon: An evaluation of artificial substrates and stocking density. *Aquaculture,* 293, 42-48.

Arnold, S.J., Sellars, M.J., Crocos, P.J. and Coman, G.J., 2006. Intensive production of juvenile tiger shrimp Penaeus monodon: an evaluation of stocking density and artificial substrates. *Aquaculture,* 261,890-896.

Asaduzzaman, M., Wahab, M.A., Verdegem, M.C.J., Huque, S., Salam, M.A. and Azim, M.E., 2008. C/N ratio control and substrate addition for periphyton development jointly enhance freshwater prawn Macrobrachium rosenbergii production in ponds. *Aquaculture,* 280,117-123.

Avnimelech, Y. 2012. Biofloc Technology - A Practical Guide Book, 2d Edition. The World Aquaculture Society, Baton Rouge, Louisiana, United States.

Avnimelech, Y. and Kochba, M., 2009. Evaluation of nitrogen uptake and excretion by tilapia in bio floc tanks, using 15N tracing. *Aquaculture,* 287,163-168.

Avnimelech, Y. and Ritvo, G., 2003. Shrimp and fish pond soils: processes and management. *Aquaculture,* 220,.549-567.

Avnimelech, Y. and Zohar, G., 1986. The effect of local anaerobic conditions on growth retardation in aquaculture systems. *Aquaculture,* 58.167-174.

Avnimelech, Y., 1999. Carbon/nitrogen ratio as a control element in aquaculture systems. *Aquaculture,* 176, 227-235.

Avnimelech, Y., 2006. Bio-filters: the need for and new comprehensive approach. *Aquacult. Eng.,* 34,172-178.

Avnimelech, Y., 2007. Feeding with microbial flocs by tilapia in minimal discharge bio-flocs technology ponds. *Aquaculture,* 264,140-147.

Avnimelech, Y., 2009. Biofloc technology. A practical guide book. The World Aquaculture Society, Baton Rouge, p.182.

Avnimelech, Y., Kochva, M. and Diab, S., 1994. Development of controlled intensive aquaculture systems with a limited water exchange and adjusted carbon to nitrogen ratio. *Isr. J. Aquacult -Bamidgeh,* 46, 119-131.

Avnimelech, Y., Mokady, S. and Schroeder, G.L., 1989. Circulated ponds as efficient bioreactors for single cell protein production. *Isr. J. Aquacult.,* 41,58-66.

Avnimelech, Y., Weber, B., Hepher, B., Milstein, A. and Zorn, M., 1986. Studies in circulated fish ponds: organic matter recycling and nitrogen transformation. *Aquacult. Res.,* 17,231-242.

Azim, M.E. and Little, D.C., 2008. The biofloc technology (BFT) in indoor tanks: Water quality, biofloc composition, and growth and welfare of Nile tilapia, *Oreochromis niloticus. Aquaculture,* 283, 29–35.

Azim, M.E., Little, D. and North, B., 2007. Growth and welfare of Nile tilapia (Oreochromis niloticus) cultured indoor tank using biofloc technology (BFT). In *Procedings of Aquaculture Conference 2007* (Vol. 26).

Barbusiński, K. and Kościelniak, H., 1995. Influence of substrate loading intensity on floc size in activated sludge process. *Water Res.,* 29,1703-1710.

Bauer, W., Prentice-Hernandez, C., Tesser, M.B., Wasielesky Jr, W. and Poersch, L.H., 2012. Substitution of fishmeal with microbial floc meal and soy protein concentrate in diets for the pacific white shrimp Litopenaeus vannamei. *Aquaculture,* 342, 112-116.

Biggs, C.A. and Lant, P.A., 2000. Activated sludge flocculation: on-line determination of floc size and the effect of shear. *Water Res.,* 342542-2550.

Boadway, J.D., 1978. Dynamics of growth and breakage of alum floc in presence of fluid shear. *J. Environ. Eng.*, 104901-915.

Boyd, C.E., Thunjai, T. and Boonyaratpalin, M., 2002. Dissolved salts in water for inland low-salinity shrimp culture. *Global Aquaculture Advocate*, 5,40-45.

Brokhoff, J. C. P. and Linsen, B. G., studies in pore systems in absorbents and catalysts, in physical chemical aspects of absorbents and catalysts, Linsen, B. G., Ed., new York 1970.

Brown, A.H.D. and Briggs, J.D., 1991. Sampling Strategies for Genetic. *Genetics and conservation of rare plants/DA Falk [et al.].–New York*, pp.99-119.

Bureau, C., carlos Garcia-Pagan, J., Otal, P., Pomier-Layrargues, G., Chabbert, V., Cortez, C., Perreault, P., marie Péron, J., Abraldes, J.G., Bouchard, L. and Bilbao, J.I., 2004. Improved clinical outcome using polytetrafluoroethylene-coated stents for TIPS: results of a randomized study. *Gastroenterology*, 126,469-475.

Bureau, D.P., Harris, A.M. and Cho, C.Y., 1999. Apparent digestibility of rendered animal protein ingredients for rainbow trout (Oncorhynchus mykiss). *Aquaculture*, 180,345-358.

Burford, M. A., Thompson, P. J., Mc Intosh, Bauman, R. H. and Pearson, D. C., 2003. Nutrient and microbial dynamics in high intensity, zero exchange shrimp ponds in Belize. *Aquaculture*, 219, 393 – 411.

Burford, M.A., Sellars, M.J., Arnold, S.J., Keys, S.J., Crocos, P.J. and Preston, N.P., 2004. Contribution of the natural biota associated with substrates to the nutritional requirements of the post-larval shrimp, Penaeus esculentus (Haswell), inhigh-density rearing systems. *Aquacult. Res.*, 35, 508-515.

Chaignon, V., Lartiges, B.S., El Samrani, A. and Mustin, C., 2002. Evolution of size distribution and transfer of mineral particles between flocs in activated sludges: an insight into floc exchange dynamics. *Water Res.*, 36, 676-684.

Chamberlain, G., Avnimelech, Y., McIntosh, R.P. and Velasco, M., 2001. Advantages of aerated microbial reuse systems with balanced C: N-I: Nutrient transformation and water quality benefits. Global aquaculture advocate, 4(2), pp.53-56.

Cheng, W., Chen, S.M., Wang, F.I., Hsu, P.I., Liu, C.H. and Chen, J.C., 2003. Effects of temperature, pH, salinity and ammonia on the phagocytic activity and clearance efficiency of giant freshwater prawn Macrobrachium rosenbergii to Lactococcus garvieae. *Aquaculture*, 219,111-121.

Cherry, I.S. and Crandall Jr, L.A., 1932. The specificity of pancreatic lipase: its appearance in the blood after pancreatic injury. *Am. J. Physiol. -Legacy Content*, 100, 266-273.

Chhaya, V. and Kumar, R., 2014. Utilization of distillery wastewater in fertigation: A beneficial use. *Int. j. res. chem. environ.*, 4,1-9.

Chien, Y.H., Pan, C.H. and Hunter, B., 2003. The resistance to physical stresses by Penaeus monodon juveniles fed diets supplemented with astaxanthin. *Aquaculture*, 216, 177-191.

Cho, C.Y., Slinger, S.J. and Bayley, H.S., 1982. Bioenergetics of salmonid fishes: energy intake, expenditure and productivity. *Comp. Biochem. Physiol., B*, 7325-41.

CLARK, J. M., 1964. Experimental Biochemistry, 228 pp. W. H. Freeman and Company, San Francisco and London.

Clifford, H.C., 1985. Semi-intensive shrimp farming. *Texas Shrimp Farming Manual. Texas Agricultural Extension Service, Corpus Christi, Texas IV*, pp.13-40.

Coackley, P. and Jones, B.R.S., 1956. Vacuum sludge filtration: I. Interpretation of results by the concept of specific resistance. *Sewage and Industrial Wastes*, 28, 963-976.

Cohen, J.M., Samocha, T.M., Fox, J.M., Gandy, R.L. and Lawrence, A.L., 2005. Characterization of water quality factors during intensive raceway production of juvenile Litopenaeus vannamei using limited discharge and biosecure management tools. *Aquacult. Eng.*, 32,425-442.

Correia, E.S., Wilkenfeld, J.S., Morris, T.C., Wei, L., Prangnell, D.I. and Samocha, T.M., 2014. Intensive nursery production of the Pacific white shrimp Litopenaeus vannamei using two commercial feeds with high and low protein content in a biofloc-dominated system. *Aquacult. Eng.,*, 59, 48-54.

Costa-Pierce, B.A. and Page, G.G., 2012. Aquaculture, Sustainability Science in. In *Encyclopedia of Sustainability Science and Technology* (pp. 564-581). Springer, New York, NY.

Crab, R., 2010. Bioflocs technology: an integrated system for the removal of nutrients and simultaneous production of feed in aquaculture (Doctoral dissertation, Ghent University).

Crab, R., Avnimelech, Y., Defoirdt, T., Bossier, P. and Verstraete, W., 2007. Nitrogen removal techniques in aquaculture for a sustainable production. *Aquaculture*, 270, 1-14.

Crab, R., Chielens, B., Wille, M., Bossier, P. and Verstraete, W., 2010. The effect of different carbon sources on the nutritional value of bioflocs, a feed for Macrobrachium rosenbergii postlarvae *Aquacult. Res.*, 41,559-567.

Crab, R., Defoirdt, T., Bossier, P and Verstraete, W., 2012. Biofloc technology in aquaculture: Beneficial effects and future challenges. *Aquaculture*, 356-357, 351–356

Crab, R., Kochva, M., Verstraete, W. and Avnimelech, Y., 2009. Bio-flocs technology application in over-wintering of tilapia. *Aquacult. Eng.,*, 40,105-112.

Crab, R., Lambert, A., Defoirdt, T., Bossier, P. and Verstraete, W., 2010. The application of bioflocs technology to protect brine shrimp (Artemia franciscana) from pathogenic Vibrio harveyi. *J. Appl Microbiol.*, 109, 1643-1649.

Craig, S., L.A. Helfrich. 2002. Understanding Fish Nutrition, Feeds and Feeding (Publication 420–256). Virginia Cooperative Extension, Yorktown (Virginia). 4.

Cuzon, G., Lawrence, A., Gaxiola, G., Rosas, C. and Guillaume, J., 2004. Nutrition of Litopenaeus vannamei reared in tanks or in ponds. *Aquaculture*, 235, 513-551.

Da Silva BC., Vieira FN., Mouriño JLP., Ferreira GS., WQ Seiffert. 2013. Salts of organic acids selection by multiple characteristics for marine shrimp nutrition. *Aquaculture* 384–387, 104-110 pp.

D'Abramo, L.R., Conklin, D.E. and Akiyama, D.M. eds., 1997. *Crustacean nutrition* (Vol. 6). World Aquaculture Society.

Davis, D.A., Lawrence, A.L. and Gatlin, D.M., 1993. Response of Penaeus vannamei to dietary calcium, phosphorus and calcium: phosphorus ratio. *J. World Aquacult. Soc.*, 24, 504-515.

De Barros, H.P. and Valenti, W.C., 2003. Food intake of Macrobrachium rosenbergii during larval development. *Aquaculture*, 216, 165-176.

De Schryver, P. and Verstraete, W., 2009. Nitrogen removal from aquaculture pond water by heterotrophic nitrogen assimilation in lab-scale sequencing batch reactors. *Bioresour Technol.*, 100, 1162-1167.

De Schryver, P., Crab, R., Defoirdt, T., Boon, N. and Verstraete, W., 2008. The basics of bio-flocs technology: the added value for aquaculture. *Aquaculture*, 277, 125-137.

Decamp, O., Conquest, L., Forster, I. and Tacon, A.G.J., 2002. The nutrition and feeding of marine shrimp within zero-water exchange aquaculture production systems: role of eukaryotic microorganisms.

Ebeling, J.M., Timmons, M.B and Bisogni, J.J.J., 2006. Engineering analysis of the stoichiometry of photoautotrophic, autotrophic, and heterotrophic removal of ammonia–nitrogen in aquaculture systems. *Aquaculture*, 257, 346–358.

Ekasari, J., Azhar, M.H., Surawidjaja, E.H., Nuryati, S., De Schryver, P. and Bossier, P., 2014. Immune response and disease resistance of shrimp fed biofloc grown on different carbon sources. *Fish Shellfish Immunol.*, 41, 332-339.

Ekasari, J., Crab, R. and Verstraete, W., 2010. Primary nutritional content of bio-flocs cultured with different organic carbon sources and salinity. HAYATI J. Biosci., 17, 125-130.

Ekasari, J., Suprayudi, M.A., Wiyoto, W., Hazanah, R.F., Lenggara, G.S., Sulistiani, R., Alkahfi, M. and Zairin Jr, M., 2016. Biofloc technology application in African catfish fingerling production: the effects on the reproductive performance of broodstock and the quality of eggs and larvae. *Aquaculture*, 464, 349-356.

Emerenciano, M., Ballester, E.L., Cavalli, R.O. and Wasielesky, W., 2012. Biofloc technology application as a food source in a limited water exchange nursery system for pink shrimp Farfantepenaeus brasiliensis (Latreille, 1817). *Aquacult. Res.*, 43, 447-457.

Emerenciano, M., Cuzon, G., Arévalo, M., Miquelajauregui, M.M. and Gaxiola, G., 2013. Effect of short-term fresh food supplementation on reproductive performance, biochemical composition, and fatty acid profile of Litopenaeus vannamei (Boone) reared under biofloc conditions. *Aquacult. Int.*, 21, 987-1007.

Esparza-Leal, H.M., Cardozo, A.P. and Wasielesky Jr, W., 2015. Performance of Litopenaeus vannamei postlarvae reared in indoor nursery tanks at high stocking density in clear-water versus biofloc system. *Aquacult. Eng.,*, 68, 28-34.

Fast, A.W., 1991. MARINE SHRIMP POND GROWOUT CONDITIONS AND STRATEGIES-A REVIEW AND PROGNOSIS. *Rev. Aquat. Sci.*, 3, 357-399.

Federation, W.E. and American Public Health Association, 2005. Standard methods for the examination of water and wastewater. American Public Health Association (APHA): Washington, DC, USA.

Felix, S., 2006. Shrimp raceway technology. DBT, New Delhi &TANUVAS, Chennai: pp; 9-15.

Felix, S., 2012. Biosecured Aerobic Flocculent driven raceway farming technology for mass production of ornamental fishes in India. Proceedings of the international conference on sustainable ornamental fisheries – way forward. Cochin University of science and technology, Kochi, 23 to 25 march, 2012. pp 200.

Felix, S., Antony Jesu Prabhu, P., Cheryl Antony, Gopalakannan, A., Rajaram, R., 2015a. Studies on nursery rearing of Genetically Improved Farmed Tilapia (GIFT) in Bio-secured raceway systems in India.J. *Indian Fish.Assoc.*, 42:1-10.

Felix, S., Cheryl Antony, Gopalakannan. A., Antony Jesu Prabhu, P., Rajaram R., 2015b.Intensive nursery rearing of *Litopenaeus vannamei* in Aerobic Microbial Floc (AMF) driven low saline raceway system. World Aquaculture 2015 - Meeting Abstract

Fisheries, F.A.O., 2016. Aquaculture Department: Rome.

Fox, J.M., Davis, D.A., Wilson, M. and Lawrence, A.L., 2006. Current status of amino acid requirement research with marine penaeid shrimp. *Avances en Nutrición Acuícola, 8*, pp.182-196.

Gamboa-delgado, J., Molina-poveda, C. and Cahu, C., 2003. Digestive enzyme activity and food ingesta in juvenile shrimp Litopenaeus vannamei (Boone, 1931) as a function of body weight. *Aquacult. Res.*, 34, 1403-1411.

Garatun-Tjeldstø, O., Otterå, H., Julshamn, K. and Austreng, E., 2006. Food ingestion in juvenile cod estimated by inert lanthanide markers–effects of food particle size. *ICES J. Mar. Sci.,* 63, 311-319.

Garczarczyk, J. J. (1996). Microbial films and microbial flocs some similarities and *differences*. Third International IAWQ Special Conference on Biofilm Systems, Copenhagen.

Ghaly, A.E., Kok, R. and Ingrahm, J.M., 1989. Growth rate determination of heterogeneous microbial population in swine manure. *Appl. Biochem. Biotechnol.*, 22, 59.

Glencross, B.D., 2009. Exploring the nutritional demand for essential fatty acids by aquaculture species. *Rev. aquacult.*, 1, 71-124.

Gross, A., Boyd, C.E. and Wood, C.W., 2000. Nitrogen transformations and balance in channel catfish ponds. *Aquacult. Eng.,*, 24, 1-14.

Halver, J.E., 1976. The nutritional requirements of cultivated warmwater and coldwater fish species. In *FAO Technical Conference on Aquaculture, Kyoto (Japan), 26 May 1976.*

Hargreaves, J., 2006. Photosynthetic suspended-growth systems in Aquaculture. *Aquacult. Eng.,* 34: 344–363.

Hargreaves, J.A., 2013. Biofloc Production Systems for Aquaculture. Southern Regional Aquaculture Centre, No. 4503.

Hari, B., Kurup, B. M ., Johny T. Varghese., Schrama, J.W .and Verdegem, M.C.J., 2006. The effect of carbohydrate addition on water quality and the nitrogen budget in extensive shrimp culture system. *Aquaculture,* 252: 248 – 263.

Hari, B., Kurup, B. M, Johny T.V, Schrama, J.W. and Verdegem, M.C.J., 2004. Improved sustainability in extensive shrimp culture system: control of carbon nitrogen ratio through addition of carbohydrate to the pond. *Aquaculture,* 241: 179 – 194.

Harris, J.M., 1993. The presence, nature, and role of gut microflora in aquatic invertebrates: a synthesis. *Microb. Ecol.*, 25, 195-231.

Hastings, W.H., 1969. Nutritional score. In *Fish in research* (pp. 263-292).

Hende, S.F.D., Claessens, L., Muylder, E.D., Boon, N., Vervaeren, H., 2014. Microbial bacterial flocs originating from aquaculture wastewater treatment as diet ingredient for *Litopenaeus vannamei* (Boone). *Aquaculture. Res.* 47, 1075 - 1089.

Hermanowicz, S.W. and Ganczarczyk, J.J., 1983. Some fluidization characteristics of biological beds. *Biotechnol. Bioeng.*, 25, 1321-1330.

Hollender, J., van der, Krol, D., Kornberger, L., Gierden, E., Dott, W., 2002.Effect of different carbon sources on the enhanced biological phosphorus removal in a sequencing batch reactor. *World J. Microbiol. Biotechnol.* 18, 355–360.

HUANG, S.J., XIE, L.W., ZHANG, M., WU, W.H. and SHEN, L.C., 2004. LIU Jie on, Chengdu University of Technology, China); Formation mechanism of authigenic chlorite and relation to preservation of porosity in nonmarine Triassic reservoir sandstones, Ordos Basin and Sichuan Basin. *China, 3.*

Izquierdo, M., Forster, I., Divakaran, S., Conquest, L., Decamp, O. and Tacon, A., 2006. Effect of green and clear water and lipid source on survival, growth and biochemical composition of Pacific white shrimp Litopenaeus vannamei. *Aquacult. Nutr.*, 12, 192-202.

Jain, R. and Srivastava, S., 2012. Nutrient composition of spent wash and its impact on sugarcane growth and biochemical attributes. Physiol. *Mol. Biol. Plants*, 18, 95-99.

Jatobá, A., da Silva, B.C., da Silva, J.S., do Nascimento Vieira, F., Mouriño, J.L.P., Seiffert, W.Q. and Toledo, T.M., 2014. Protein levels for Litopenaeus vannamei in semi-intensive and biofloc systems. *Aquaculture*, 432, 365-371.

Jatobá, A., da Silva, B.C., da Silva, J.S., do Nascimento Vieira, F., Mouriño, J.L.P., Seiffert, W.Q. and Toledo, T.M., 2014. Protein levels for Litopenaeus vannamei in semi-intensive and biofloc systems. *Aquaculture*, 432, 365-371.

Ju, Z.Y., Forster, I., Conquest, L. and Dominy, W., 2008. Enhanced growth effects on shrimp (Litopenaeus vannamei) from inclusion of whole shrimp floc or floc fractions to a formulated diet. *Aquacult. Nut.*, 14, 533-543.

Ju, Z.Y., Forster, I., Conquest, L., Dominy, W., Kuo, W.C. and David Horgen, F., 2008b. Determination of microbial community structures of shrimp floc cultures by biomarkers and analysis of floc amino acid profiles. *Aquacult. Res.*, 39, 118-133.

Keiding, K. and Nielsen, P.H., 1997. Desorption of organic macromolecules from activated sludge: effect of ionic composition *Water Res.*, 31, 1665-1672.

Kiessling, A. and Askbrandt, S., 1993. Nutritive value of two bacterial strains of single-cell protein for rainbow trout (Oncorhynchus mykiss). *Aquaculture*, 109, 119-130.

Kim, S.K., Pang, Z., Seo, H.C., Cho, Y.R., Samocha, T. and Jang, I.K., 2014. Effect of bioflocs on growth and immune activity of Pacific white shrimp, Litopenaeus vannamei postlarvae. *Aquacult. Res.*, 45, 362-371.

Knights, B., 1983. Food particle-size preferences and feeding behaviour in warmwater aquaculture of European eel, Anguilla anguilla (L.). *Aquaculture*, 30, 173-190.

Krummenauer, D., Peixoto, S., Cavalli, R.O., Poersch, L.H. and Wasielesky, W., 2011. Superintensive culture of white shrimp, Litopenaeus vannamei, in a biofloc technology system in southern Brazil at different stocking densities. *J. World Aquacult. Soc.*, 42, 726-733.

Krummenauer, D., Poersch, L., Romano, L.A., Lara, G.R., Encarnação, P. and Wasielesky Jr, W., 2014. The effect of probiotics in a Litopenaeus vannamei biofloc culture system infected with Vibrio parahaemolyticus. *J. Appl. Aquacult.*, 26,370-379.

Kuhn, D.D., Boardman, G.D., Lawrence, A.L., Marsh, L. and Flick Jr, G.J., 2009. Microbial floc meal as a replacement ingredient for fish meal and soybean protein in shrimp feed. *Aquaculture*, 296, 51-57.

Kuhn, D.D., Lawrence, A.L., Boardman, G.D., Patnaik, S., Marsh, L. and Flick Jr, G.J., 2010. Evaluation of two types of bioflocs derived from biological treatment of fish effluent as feed ingredients for Pacific white shrimp, Litopenaeus vannamei. *Aquaculture*, 303, 28-33.

Kumar, V., Wati, L., Nigam, P., Banat, I.M., Yadav, B.S., Singh, D. and Marchant, R., 1998. Decolorization and biodegradation of anaerobically digested sugarcane molasses spent wash effluent from biomethanation plants by white-rot fungi. *Process Biochem.*, 33, 83-88.

Lancelot, C. and Billen, G., 1985. Carbon-nitrogen relationship in nutrient metabolism of coastal marine ecosystem. *Advances in Aquatic Microbiology,* 1, 263-321.

Lara-Flores, M., Olvera-Novoa, M.A., Guzmán-Méndez, B.E. and López-Madrid, W., 2003. Use of the bacteria Streptococcus faecium and Lactobacillus acidophilus, and the yeast Saccharomyces cerevisiae as growth promoters in Nile tilapia (Oreochromis niloticus). *Aquaculture,* 216, 193-201.

Lee, D.J., Chen, G.W., Liao, Y.C. and Hsieh, C.C., 1996. On the free-settling test for estimating activated sludge floc density. *Water Res.,* 30, 541-550.

Lee, P.G., Smith, L.L. and Lawrence, A.L., 1984. Digestive proteases of Penaeus vannamei Boone: relationship between enzyme activity, size and diet. *Aquaculture,* 42, 225-239.

Li, D.H. and Ganczarczyk, J.J., 1986. Application of image analysis system for study of activated sludge flocs. *Water Pollut. Control Res. J. Can.,* 21, 130-140.

Li, D.H., Ganczarczyk, J.J. and Jenkins, D., 1986. Physical characteristics of activated sludge flocs. *Crit. Rev. Environ. Sci.* Technol. , 17, 53-87.

Lim, C., Ako, H., Brown, C.L. and Hahn, K., 1997. Growth response and fatty acid composition of juvenile Penaeus vannamei fed different sources of dietary lipid. *Aquaculture,* 151, 143-153.

Liu, W.C., Luo, G.Z., Li, L., Wang, X.Y., Wang, J., Ma, N.N., Sun, D.C. and Tan, H.X., 2017. Nitrogen Dynamics and Biofloc Composition using Biofloc Technology to Treat Aquaculture Solid Waste Mixed with Distillery Spent Wash. *N. Am. J. Aquac.,* 79, 27-35.

Liu, Y. and Fang, H.H., 2003. Influences of extracellular polymeric substances (EPS) on flocculation, settling, and dewatering of activated sludge.

López-Elías, J.A., Moreno-Arias, A., Miranda-Baeza, A., Martínez-Córdova, L.R., Rivas-Vega, M.E. and Márquez-Ríos, E., 2015. Proximate composition of bioflocs in culture systems containing hybrid red tilapia fed diets with varying levels of vegetable meal inclusion. *N. Am. J. Aquac.*77, 102-109.

Lu, L., Tan, H., Luo, G. and Liang, W., 2012. The effects of Bacillus subtilis on nitrogen recycling from aquaculture solid waste using heterotrophic nitrogen assimilation in sequencing batch reactors. Bioresour. Technol., 124, 180-185.

Luo, G., Gao, Q., Wang, C., Liu, W., Sun, D., Li, L. and Tan, H., 2014. Growth, digestive activity, welfare, and partial cost-effectiveness of genetically improved farmed tilapia (Oreochromis niloticus) cultured in a recirculating aquaculture system and an indoor biofloc system. *Aquaculture,* 422, 1-7.

Luo, G.Z., Avnimelech, Y., Pan, Y.F. and Tan, H.X., 2013. Inorganic nitrogen dynamics in sequencing batch reactors using biofloc technology to treat aquaculture sludge. *Aquacult. Eng.,* 52, 73-79.

Magara, Y., Nambu, S. and Utosawa, K., 1976. Biochemical and physical properties of an activated sludge on settling characteristics. *Water Res,* 10, .71-77.

Mahmoud, N.S. and Ghaly, A.E., 2004. Influence of temperature and pH on the nonenzymatic reduction of triphenyltetrazolium chloride. *Biotechnol. Prog.*, 20, 346-353.

Maicá, P.F., de Borba, M.R. and Wasielesky Jr, W., 2012. Effect of low salinity on microbial floc composition and performance of *Litopenaeus vannamei* (Boone) juveniles reared in a zero-water-exchange super-intensive system. *Aquacult. Res.*, 43, 361-370.

Martins, A.M.P., Heijnen, J.J. and Van Loosdrecht, M.C.M., 2003. Effect of dissolved oxygen concentration on sludge settleability. *Appl. Microbiol. Biotechnol.*, *62*(5-6), pp.586-593.

Mc Intosh, R.P., 2000. Changing paradigms in shrimp farming IV , low protein feeds and feeding strategies, *Global Aquaculture Advocate* 3 (2): 44-50.

McCarty, P.L. and Rittmann, B.E., 2001. Environmental biotechnology: principles and applications. *Ed. McGraw Mil. Cap*, *2*, pp.19-55.

McGraw, W., Teichert-Coddington, D.R., Rouse, D.B. and Boyd, C.E., 2001. Higher minimum dissolved oxygen concentrations increase penaeid shrimp yields in earthen ponds. *Aquaculture*, 199, 311-321.

McGraw, W.J. and Scarpa, J., 2003. Minimum environmental potassium for survival of Pacific white shrimp Litopenaeus vannamei (Boone) in freshwater. *J. shellfish* ***res***, 22, 263-268.

McIntosh, D., Samocha, T.M., Jones, E.R., Lawrence, A.L., McKee, D.A., Horowitz, S. and Horowitz, A., 2000. The effect of a commercial bacterial supplement on the high-density culturing of Litopenaeus vannamei with a low-protein diet in an outdoor tank system and no water exchange. *Aquacult. Eng.*, 21, 215-227.

McSwain, B.S., Irvine, R.L., Hausner, M. and Wilderer, P.A., 2005. Composition and distribution of extracellular polymeric substances in aerobic flocs and granular sludge. *Appl. Environ. Microbiol.*, 71, 1051-1057.

McWhinnie, M.A. and Kirchenberg, R.J., 1962. Crayfish hepatopancreas metabolism and the intermoult cycle. *Comp. Biochem. Physiol.*, 6, 117-128.

Menaga, M. and Felix, S., 2015. GROWTH OF GENETICALLY IMPROVED FARMED TILAPIA (OREOCHROMIS NILOTICUS) IN INDOOR RECIRCULATING AQUACULTURE SYSTEM.

Menaga, M., Felix, S., Gopalakannan, A., 2017. Distillery wastage (spent wash) as a novel carbon source for aquaculture intensification. Indian Vet J. 94(12), 15-17.

Mente, E., Coutteau, P., Houlihan, D., Davidson, I. and Sorgeloos, P., 2002. Protein turnover, amino acid profile and amino acid flux in juvenile shrimp Litopenaeus vannamei: effects of dietary protein source. *J. Exp. Biol.*, 205, 3107-3122.

Meyers, S.P. and Latscha, T., 1997. Carotenoids. In advances in world aquaculture. Vo. 6. Crustacean Nutrition. *World Aquaculture Society*.

Mikkelsen, L.H., Gotfredsen, A.K., Agerbaek, M.L., Nielsen, P.H. and Keiding, K., 1996. Effects of colloidal stability on clarification and dewatering of activated sludge. *Water Sci. Technol.* , 34, 449-457.

Mitani, T., 1983. A structure model of activated sludge flocs. *Suishitsu Odaku Kenkyuu, 6,* pp.69-77.

Moreno-Arias, A., López-Elías, J.A., Martínez-Córdova, L.R., Ramírez-Suárez, J.C., Carvallo-Ruiz, M.G., García-Sánchez, G., Lugo-Sánchez, M.E. and Miranda-Baeza, A., 2018. Effect of fishmeal replacement with a vegetable protein mixture on the amino acid and fatty acid profiles of diets, biofloc and shrimp cultured in BFT system. *Aquaculture,* 483, 53-62.

Moss, R.E., Seed, R.B., Kayen, R.E., Stewart, J.P., Der Kiureghian, A. and Cetin, K.O., 2006. CPT-based probabilistic and deterministic assessment of in situ seismic soil liquefaction potential. *J. Geotech. Geoenviron.* , 132, 1032-1051.

Moss, S.M., Arce, S.M., Argue, B.J., Otoshi, C.A., Calderon, F.R. and Tacon, A.G., 2001. Greening of the blue revolution: efforts toward environmentally responsible shrimp culture. In *The New Wave, Proceedings of the special session on sustainable shrimp culture, aquaculture* (1-19).

Moss, S.M., LeaMaster, B.R. and Sweeney, J.N., 2000. Relative Abundance and Species Composition of Gram-Negative, Aerobic Bacteria Associated with the Gut of Juvenile White Shrimp Litopenaeus vannamei Reared in Oligotrophic Well Water and Eutrophic Pond Water. *J. World Aquacult. Soc.,* 31, 255-263.

National Research Council, 2011. *Nutrient requirements of fish and shrimp*. National academies press.

Nunes, A.J., Castro, L.F. and Sabry-Neto, H., 2010. Microbial flocs spare protein in white shrimp diets. *Global aquaculture advocate.*

Nunes, A.J., Sá, M.V., Andriola-Neto, F.F. and Lemos, D., 2006. Behavioral response to selected feed attractants and stimulants in Pacific white shrimp, Litopenaeus vannamei. *Aquaculture,* 260, 244-254.

Oehmen, A., Yuan, Z., Blackall, L.L. and Keller, J., 2004. Short-term effects of carbon source on the competition of polyphosphate accumulating organisms and glycogen accumulating organisms. *Water Sci. Technol.*, 50, 139-144.

Otoshi, C.A., Tang, L.R., Moss, D.R., Arce, S.M., Holl, C.M. and Moss, S.M., 2009. Performance of Pacific white shrimp, Penaeus (Litopenaeus) vannamei, cultured in biosecure, super-intensive, recirculating aquaculture systems. *BROWDY, CL; JORY, DE The rising tideproceedings of the special session on sustainable shrimp farming. Baton Rouge: World Aquaculture Society.*

Pan, Y., Qiu, J. and Silverman, R.B., 2003. Design, synthesis, and biological activity of a difluoro-substituted, conformationally rigid vigabatrin analogue as a potent γ-aminobutyric acid aminotransferase inhibitor. *J. Med. Chem.,* 46, 5292-5293.

Panjaitan, P., 2004. *Field and laboratory study of Penaeus monodon culture with zero water exchange and limited water exchange model using molasses as a carbon source*(Doctoral dissertation, Ph. D. Thesis, Charles Darwin Univ., Darwin, NT, Australia).

Peñaflorida, V.D., 1989. An evaluation of indigenous protein sources as potential component in the diet formulation for tiger prawn, Penaeus monodon, using essential amino acid index (EAAI). *Aquaculture*, 83, 319-330.

Perez-Fuentes, J.A., Hernández-Vergara, M.P., Perez-Rostro, C.I. and Fogel, I., 2016. C: N ratios affect nitrogen removal and production of Nile tilapia Oreochromis niloticus raised in a biofloc system under high density cultivation. *Aquaculture*, 452, 247-251.

Perez-Fuentes, J.A., Pérez-Rostro, C.I. and Hernández-Vergara, M.P., 2013. Pond-reared Malaysian prawn Macrobrachium rosenbergii with the biofloc system. *Aquaculture*, 400, pp.105-110.

Peterson, J.J. and Griffith, D.R.W., 1999. Intensive nursery systems. *Global Aquaculture Advocate*, 2, 60-61.

Prangnell, D.I., Castro, L.F., Ali, A.S., Browdy, C.L., Zimba, P.V., Laramore, S.E. and Samocha, T.M., 2016. Some Limiting Factors in Superintensive Production of Juvenile Pacific White Shrimp, Litopenaeus vannamei, in No-water-exchange, Biofloc-dominated Systems. *J. World Aquacult. Soc.*, 47, 396-413.

Raj, A., Menaga, M. and Felix, S., EFFECT OF C/N RATIO ON WATER QUALITY USING DISTILLERY SPENTWASH AS A CARBON SOURCE IN INDOOR AEROBIC MICROBIAL FLOC TANKS.

Rinchard, J., Czesny, S. and Dabrowski, K., 2007. Influence of lipid class and fatty acid deficiency on survival, growth, and fatty acid composition in rainbow trout juveniles. *Aquaculture*, 264, 363-371.

Ritvo, G., Dassa, O. and Kochba, M., 2003. Salinity and pH effect on the colloidal properties of suspended particles in super intensive aquaculture systems. *Aquaculture*, *218*(1-4), pp.379-386. Crab, R., Defoirdt, T., Bossier, P and Verstraete, W., 2012. Biofloc technology in aquaculture: Beneficial effects and future challenges. *Aquaculture*, 356-357: 351–356.

Sabry Neto, H., Santaella, S.T. and Nunes, A.J.P., 2015. Bioavailability of crude protein and lipid from biofloc meals produced in an activated sludge system for white shrimp, Litopenaeus vannamei. *R. Bras. Zootec.* , 44, 269-275.

Samocha, T. M., Lawrence, A. L., Collins, C. A., Castille, F. L., Bray, W. A., Davies, C. J. and Wood, G. F. 2004. Production of the Pacific white shrimp, *Litopenaeus vannamei*, in high-density greenhouse-enclosed raceways using low salinity groundwater. *J. Appl. Aquacult.*, 15 1-19.

Samocha, T.M. and Benner, R., 2001. Use of intensive nursery systems in commercial shrimp production operation in Texas. *Ind. Briefs*, 7, 1-3.

Samocha, T.M. and Lawrence, A.L., 1992. Shrimp nursery systems and management. *Wyban J. Proceedings of the Special Session on Shrimp Farming. USA: World Aquaculture Society. hlm, 88.*

Samocha, T.M. Wilkenfeld, J.S. Morris, T.C. Correia, E.S. and Hanson, T., 2010. Intensive raceways without water exchange analyzed for white shrimp culture. *Global Aquaculture Advocate*, 13: 22-14

Samocha, T.M., Davis, A.D., Lawrence, A.L., Collins, C.R. and Van Wyk, P., 2001. Intensive and super-intensive production of the Pacific white Litopenaeus vannamei in greenhouse-enclosed raceway systems. Book of Abstracts. *Aquaculture, 573.*

Samocha, T.M., Patnaik, S., Speed, M., Ali, A.M., Burger, J.M., Almeida, R.V., Ayub, Z., Harisanto, M., Horowitz, A. and Brock, D.L., 2007. Use of molasses as carbon source in limited discharge nursery and grow-out systems for Litopenaeus vannamei. *Aquac Eng.*, 36, 184-191.

Sandifer, P.A. and Smith, T.I., 1977. Intensive rearing of postlarval Malaysian prawns (Macrobrachium rosenbergii) in a closed cycle nursery system. *J. World Aquacult. Soc.* , 8, 225-235.

Sargent, J.R., Tocher, D.R. and Bell, J.G., 2002. The Lipids in Fish Nutrition 3rd edn,(eds Halver JE & Hardy RW) Ch. 4, 181–257.

Schneider, W.C., 1957. [99] Determination of nucleic acids in tissues by pentose analysis.

Scuras, S., Daigger, G.T. and Grady, C.L., 1998. Modeling the activated sludge floc microenvironment. *Water Sci. Technol.*, 37, 243-250.

Serra, F.P., Gaona, C.A., Furtado, P.S., Poersch, L.H. and Wasielesky, W., 2015. Use of different carbon sources for the biofloc system adopted during the nursery and grow-out culture of Litopenaeus vannamei. *Aquacult. Int.*, 23, 1325-1339.

Sinha, A.K., Baruah, K.A.R.T.I.K. and Bossier, P.E.T.E.R., 2008. Horizon scanning: the potential use of biofloc as an anti-infective strategy in aquaculture-an overview. *Aquaculture Health International, 13,* 8-10.

Soares, R., Jackson, C., Coman, F. and Preston, N., 2004 Nutritional composition of flocculated material in experimental zero-exchange system for *Penaeus monodon*. In: Proceedings of Australian *Aquaculture,* 2004, WAS, Sydney pp.89

Spicer, P.T. and Pratsinis, S.E., 1996. Shear-induced flocculation: the evolution of floc structure and the shape of the size distribution at steady state. *Water Res.*, 30, 1049-1056.

Stunner, L.N., Samocha, T.M. and Lawrence, A.L., 1992. Intensification of penaeid nursery systems. In *Marine Shrimp Culture* (pp. 321-344).

Sung, H.H., Hsu, S.F., Chen, C.K., Ting, Y.Y. and Chao, W.L., 2001. Relationships between disease outbreak in cultured tiger shrimp (Penaeus monodon) and the composition of Vibrio communities in pond water and shrimp hepatopancreas during cultivation. *Aquaculture,* 192, 101-110.

Tacon, A.G.J., Cody, J.J., Conquest, L.D., Divakaran, S., Forster, I.P. and Decamp, O.E., 2002. Effect of culture system on the nutrition and growth performance of Pacific white shrimp Litopenaeus vannamei (Boone) fed different diets. *Aquaculture nutrition*, 8, 121-137.

Tambo, N. and Watanabe, Y., 1979. Physical characteristics of flocs—I. The floc density function and aluminium floc. *Water Res.*, 13, 409-419.

Taw, N., 2006. Shrimp production in ASP system, CP Indonesia: Development of the technology from R&D to commercial production. *Aquaculture America*.

Tay, J.H., Liu, Q.S. and Liu, Y., 2001. Microscopic observation of aerobic granulation in sequential aerobic sludge blanket reactor. *J. Appl. Microbiol.*, 91, 168-175.

Tewary, A. and Patra, B.C., 2008. Use of vitamin C as an immunostimulant. Effect on growth, nutritional quality, and immune response of Labeo rohita (Ham.). *Fish Physiol Biochem.*, 34, 251-259.

Uhlířová, E., Elhottova, D., Tříska, J. and Šantrůčková, H., 2005. Physiology and microbial community structure in soil at extreme water content. Folia Microbiol. , 50, 161.

Valle, B.C.S., Dantas, E.M., Silva, J.F.X., Bezerra, R.S., Correia, E.S., Peixoto, S.R.M. and Soares, R.B., 2015. Replacement of fishmeal by fish protein hydrolysate and biofloc in the diets of Litopenaeus vannamei postlarvae. *Aquacult. Nutr.*, 21, 105-112.

Van Den Hende, S., Claessens, L., De Muylder, E., Boon, N. and Vervaeren, H., 2016. Microalgal bacterial flocs originating from aquaculture wastewater treatment as diet ingredient for Litopenaeus vannamei (Boone). *Aquac. Res.*, 47, 1075-1089.

Van Wyk P and Scarpa, J., 1999. Water quality requirements and management In: Van Wyck P (ed) Farming marine shrimp in recirculating freshwater systems. Florida Department of Agriculture and Consumer Services, Tallahasee, pp. 128–138.

Verdegem, P. and Fuchs, C., 2013. Towards a Participatory, Co-operative and Sustainable Information Society?. *Nordicom Review*, 34, 3-18.

Wall, A. and Heiskanen, J., 2003. Water-retention characteristics and related physical properties of soil on afforested agricultural land in Finland. *For. Ecol. Manage.* 186, 21-32.

Wang, G., Yu, E., Xie, J., Yu, D., Li, Z., Luo, W., Qiu, L. and Zheng, Z., 2015. Effect of C/N ratio on water quality in zero-water exchange tanks and the biofloc supplementation in feed on the growth performance of crucian carp, Carassius auratus. *Aquaculture*, 443,98-104.

Wasielesky Jr, W., Atwood, H., Stokes, A. and Browdy, C.L., 2006. Effect of natural production in a zero exchange suspended microbial floc based super-intensive culture system for white shrimp Litopenaeus vannamei. *Aquaculture*, 258, 396-403.

Webster, C.D. and Lim, C. eds., 2002. *Nutrient requirements and feeding of finfish for aquaculture*. Cabi.

Wilder, I.B. and Stanley, J.G., 1983. RNA-DNA ratio as an index to growth in salmonid fishes in the laboratory and in streams contaminated by carbaryl. *J. Fish Biol.*, 22, 165-172.

Xu, W.J. and Pan, L.Q., 2012. Effects of bioflocs on growth performance, digestive enzyme activity and body composition of juvenile Litopenaeus vannamei in zero-water exchange tanks manipulating C/N ratio in feed. *Aquaculture*, 356,147-152.

Xu, W.J. and Pan, L.Q., 2013. Enhancement of immune response and antioxidant status of Litopenaeus vannamei juvenile in biofloc-based culture tanks manipulating high C/N ratio of feed input. *Aquaculture*, 412,117-124.

Xu, W.J., Pan, L.Q., Zhao, D.H. and Huang, J., 2012. Preliminary investigation into the contribution of bioflocs on protein nutrition of Litopenaeus vannamei fed with different dietary protein levels in zero-water exchange culture tanks. *Aquaculture*, 350, 147-153.

Yao, C., Tan, H.X., Luo, G.Z. and Li, L., 2013. Effects of temperature on inorganic nitrogen dynamics in sequencing batch reactors using biofloc technology to treat aquaculture sludge. *N. Am. J. Aquac.*, 75, 463-467.

Yu, M.C., Li, Z.J., Lin, H.Z., Wen, G.L. and Ma, S., 2008. Effects of dietary Bacillus and medicinal herbs on the growth, digestive enzyme activity, and serum biochemical parameters of the shrimp Litopenaeus vannamei. *Aquacult. Int.*, 16, 471-480.

Zhao, Y., Zhang, W., Xu, W., Mai, K., Zhang, Y. and Liufu, Z., 2012. Effects of potential probiotic Bacillus subtilis T13 on growth, immunity and disease resistance against Vibrio splendidus infection in juvenile sea cucumber Apostichopus japonicas. *Fish Shellfish Immunol.*, 32, 750-755.

Zhukova, N.V. and Kharlamenko, V.I., 1999. Sources of essential fatty acids in the marine microbial loop. *Aquat. Microb. Ecol.* , 17, 153-157.

Ziaei-Nejad, S., Rezaei, M.H., Takami, G.A., Lovett, D.L., Mirvaghefi, A.R. and Shakouri, M., 2006. The effect of Bacillus spp. bacteria used as probiotics on digestive enzyme activity, survival and growth in the Indian white shrimp Fenneropenaeus indicus. *Aquaculture*, 252, 516-524.

Zita, A. and Hermansson, M., 1994. Effects of ionic strength on bacterial adhesion and stability of flocs in a wastewater activated sludge system. *J. Appl. Environ. Microbiol.*, 60, 3041-3048.

www.ingramcontent.com/pod-product-compliance
Lightning Source LLC
Chambersburg PA
CBHW062057160726
48282CB00003B/29
* 9 7 8 9 3 5 4 6 1 6 2 7 3 *